AF294041

The Reconstruction of Natural Zeolites

The Reconstruction of Natural Zeolites

by

HABBIB GHOBARKAR

Freie Universität Berlin, Institut für Mineralogie,
Berlin, Germany

and

OLIVER SCHÄF

YVAN MASSIANI

PHILIPPE KNAUTH

MADIREL, UMR6121 Université de Provence-CNRS,
Marseille, France

SPRINGER SCIENCE+BUSINESS MEDIA, B.V.

A C.I.P. Catalogue record for this book is available from the Library of Congress.

ISBN 978-1-4613-4807-8   ISBN 978-1-4419-9142-3 (eBook)
DOI 10.1007/978-1-4419-9142-3

Printed on acid-free paper

PREFACE

More than seventeen years have passed now since Glauco Gottardi and Ermano Galli [1] have published their remarkable book on « NATURAL ZEOLITES » where properties and features of naturally occuring phases then available have been compiled.
Several new natural zeolites have been found since then, but also natural counterparts of zeolites which have only been known as synthesis products. The natural formation conditions of zeolites could only be deduced and estimated from their geological environment at the time when NATURAL ZEOLITES has been published, as zeolite synthesis was mainly focused on procedures at low pressures such as those introduced by Barrer and co-workers[2]. Natural zeolites, however, had only been obtained "occasionally" and systematic study to reconstruct these formation conditions has not been performed ever since.

This book is focused on the synthesis of natural zeolites by simulating the natural synthesis conditions in the laboratory which are essentially different in means and results from those obtained by conventional synthesis methods. Although the synthesis in the laboratory has undoubtly a great number of advantages over nature such as the employment of proper precursors or the choice of pressure and temperature in a wide range, the synthesis time is very limited in respect to natural conditions: synthesis times of years or even tens of years which would be necessary to obtain synthesis results for some zeolites- e.g. at 4°C (deep sea conditions)- are rather unrealistic.

The simulation of natural formation conditions of zeolites, however, does not only allow the systematic synthesis of all natural zeolites and, thereby, the determination of optimum temperature and pressure conditions of their natural formation. The method also permits to replace the cations of the basic tetrahedral building units during the synthesis as well as the extraframework cations necessary for charge compensation. This establishes a direct precursor-product correlation with zeolites "à la carte" having the desired physical and chemical properties for new and challenging applications of the future.

This book is intended to fill this gap by tabulating not only important crystallographic, physical and chemical parameters of natural and synthetic zeolites, but also optimal synthesis parameters, especially for the hydrothermal method under high pressure. It should be a precious practical guide and tool for solid-state chemists, physicists, mineralogists and engineers. Furthermore, the structural images should convey an impression of the beauty and attractivity of this fascinating domain of advanced inorganic materials for high technology.

Berlin and Marseille, in April 2003.

The authors.

CONTENTS

1. INTRODUCTION

The synthesis of zeolites has been an ongoing subject for many years, since their first description by Cronstedt[3] in 1756. The name, derived from the Greek ζειν and λιδος means "boiling stone" showing that from the beginning the researchers were aware of their water desorption properties at elevated temperature.

Most of the natural zeolites were found to be alumino-silicates. Tetrahedral building units of aluminum and silicon are forming a three-dimensional open framework with water and cations in well defined extraframework positions localized in structure inherent channels and voids. An $[AlO_4]^{5-}$ tetrahedron, present in the $[SiO_4]^{4-}$ framework carries one excess negative charge which has to be counter balanced by the charge of either one alkali-ion or half an alkaline earth cation.

The aluminum concentration does not exceed that of silicon in principle, because, according to the Loewenstein rule, no $[AlO_4]$ tetrahedron must have another $[AlO_4]$ tetrahedron as a next neighbour. In addition, the water molecules are present in the extraframework void system of the zeolites, bound under specific, "zeolitic" conditions where they compensate the existing polar momentums on the one hand and form hydration spheres on the cations on the other hand, depending on charge distribution within the framework (given by the Al-distribution) and channel size of the respective zeolite system.

Besides alumino-silicate materials with zeolitic properties, the AlPOs do exist, showing a three-dimensional framework of $[AlO_4]^{5-}$ and $[PO_4]^{3-}$ tetrahedra, leading to an internal charge compensation of the framework without net-charges to be compensated by extraframework cations.

In alumino-silicate zeolites different grades of framework cation replacements can be observed, both on Al as well as on Si sites.

Microporous zeolite-type materials, are crystalline phases with channel diameters between 0.02 and 2nm, while, according to the IUPAC[4] nomenclature, mesoporous materials have channels in the range between 2 and 50nm. Mesoporous materials known today do not have crystalline properties like the zeolites as their framework, even if not glassy like and with well defined cristalline properties, is not charged. In consequence they do not show the same specific physical and chemical properties although bearing a considerable void system.

Macroporous materials are those showing pores bigger than 50nm.

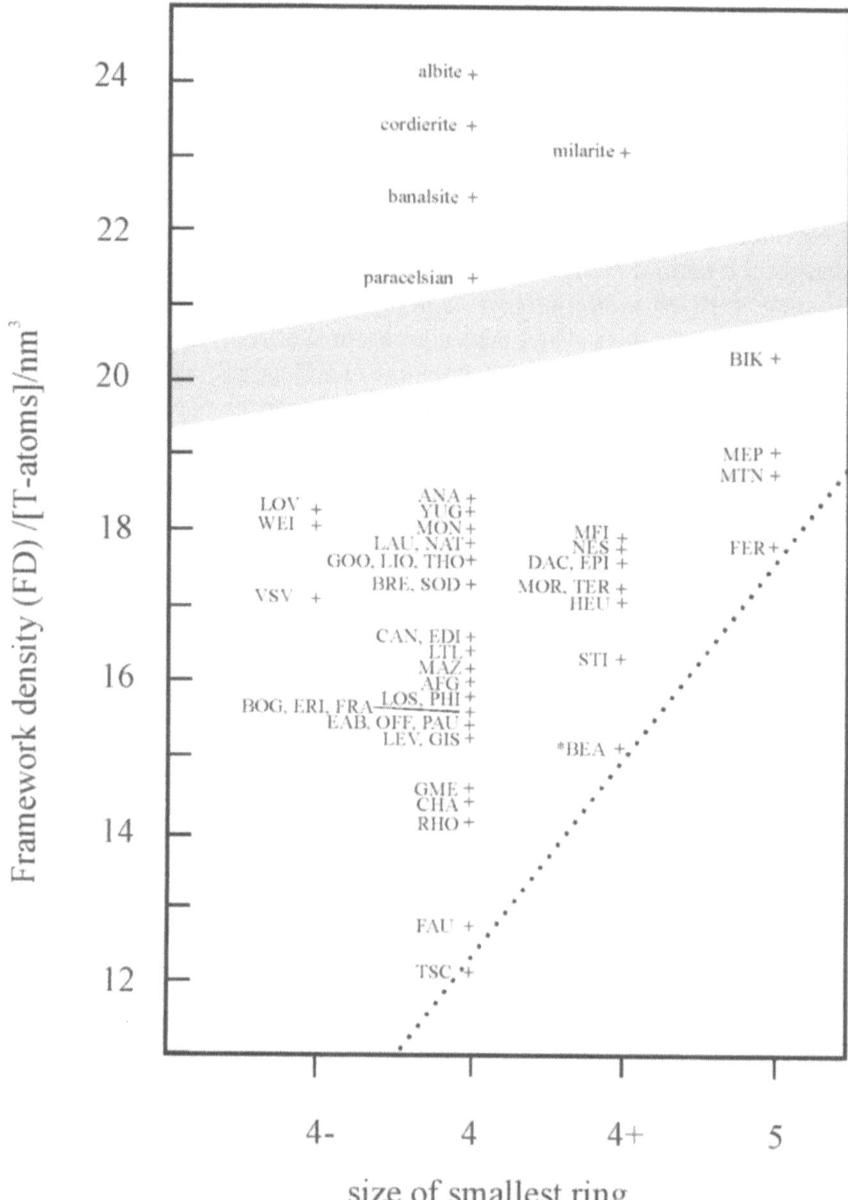

Figure 1.1: The framework density of natural zeolites versus smallest ring in loop configuration of the tetrahedras (reproduced in parts from ATLAS OF ZEOLITE FRAMEWORK TYPES [5], with kind permission).

The subcommitee on zeolites of the International Mineralogical Association, Commission on New Minerals and Mineral Names [6], gives the following definition for naturally occuring zeolite phases :

« A zeolite mineral is a crystalline substance with a structure characterized by a framework of linked tetrahedra, each consisting of four O atoms surrounding a cation. This framework contains open cavities in the form of channels and cages. These are usually occupied by H_2O molecules and extra-framework cations that are commonly exchangeable. The channels are large enough to allow the passage of guest species. In the hydrated phases, dehydration occurs at temperatures mostly below about 400 °C and is largely reversible. The framework may be interrupted by (OH,F) groups; these occupy a tetrahedron apex that is not shared with adjacent tetrahedra. »

The first basic principle for zeolites is, therefore, the existence of tetrahedral building units for the framework.

The second is the presence of a cavity system (micro-porosity) or, generally speaking, a crystallographic framework density below a certain level (as can be seen in figure 1), given that channels and voids in framework silicates considerably reduce their density. Above a certain level of framework density only conventional tectosilicates exist. Although some of the conventional tectosilicates do have channels and voids, the specific zeolite property of charged framework and charge compensating cations in the void system is not observed.

A framework density below a critical level is, however, not automatically adequate for a material to be a proper zeolite, as zeolites which consist only of cage assemblages, such as sodalite (SOD), have small framework densities but lack of the necessary channel system to show common zeolite characteristics such as adsorption of medium sized molecules and ion exchange properties to a large extent.

The third and forth characteristic principle, not strictly applicable in all cases, is the presence of water molecules and charge compensating cations in the present void system. Both are placed on well defined crystallographic positions. However, zeolites might be partially or completely dehydrated due to their natural conservation conditions or the storage conditions in the laboratory. De-alumination of the alumino-silicate material leads to simultaneous liberation of the charge compensating cations and, in consequence, also to a kind of de-hydration. In some cases high silica zeolites with low alumina content (acid zeolites, Si/Al>10) are synthesized in the laboratory and even nearly pure silica zeolites have been obtained with almost no charge compensating cations in the void system.

In contrast to alumino-silicate zeolite, common hydroxo-silicates do not fulfill the zeolite definition: given that OH groups are an integral part of their structure, their remove leads to the irreversible destruction of the lattice, which is not the case with zeolites.

Zeolite phases are distinct by the different zeolite framework types, named by a three letter code. The term framework type refers to the kind of corner-sharing network of three-dimensionally tetrahedrally coordinated atoms. In contrast, the term structure type implies both, the framework and the extra-framework constituents of a zeolite, namely the charge compensating cations, which are not considered in this classification [5].

The special method developed and applied to reconstruct natural zeolites, also allows the synthesis of zeolites with different symmetries and morphologies, dependent on the zeolite framework type under investigation. Synthesized natural zeolites are shown according to the most abundant morphology directly related to the symmetry (example: representation of monoclinic Wairakite instead of cubic).

Synthesized natural zeolites do have special properties compared to their natural counterparts. They are « fresh » materials with respect to the original, maximum water content of zeolites at the moment of their formation. Synthetic natural zeolites are also very pure as they only consist of components given by the precursor composition. In contrast, natural zeolites are very often polluted: a very eloquent example is the secondary ion exchange of zeolites formed by alteration of volcanic rocks which subsequently came into contract with sea water. This is an example where the ion exchange process is comprehensible, however, in many other cases such processes can only be assumed.

There are also strong hints that zeolites synthesized under conventional laboratory conditions are showing big differences in physical and chemical properties with respect to their natural counterparts: the synthesis method introduced in this book uses glass precursors of the respective zeolite composition which are transformed to the desired zeolite under high water pressure at elevated temperatures at extended times. Natural zeolites could be sythesized in this way, not obtainable by other methods confirming that it is a of simulation of the natural formation conditions. The submission of natural zeolites under these conditions also does not alter their properties even after prolonged time. However, submitting conventional synthesized zeolites with the same framework type to these conditions leads to their transformation to other zeolites[7].

Zeolite phases obtained by the synthesis method simulating their natural formation conditions are phases stable for kinetic, rather than for thermodynamic reasons. One example, zeolite ABW (zeolite A- Barrer, White), can be synthesized in the laboratory, but is not found in nature. In nature only eucryptite, a phase with the same composition, but higher framework density, is stable (see also chapter 3.2). The lack of thermodynamic stability can be deduced from the fact that starting from a glassy precursor, at a given pressure and time of synthesis, with rising synthesis temperature ABW is increasingly substituted by eucryptite[8]. At present, the hypothesis is established that all zeolite phases might just be kinetically favoured phases rather than the denser thermodynamically stable tectosilicates, but remains unconfirmed.

The occurrences and localities where natural zeolites can be found are not mentioned in this book, given that the zeolites presented are all synthesized and, therefore, independent from formation conditions in nature. A comprehensive compilation of natural occurrences exists in the book of Gottardi and Galli[1].

The representation of zeolite crystal structures, illustrations of channel systems as well as pictures of zeolite cages are done in a developing way, so that the reader gets more and more familiar with this topic containing some of the most complicated inorganic structures known today.

2. GENERAL DESCRIPTION OF THE ZEOLITES

2.1 Grouping of Zeolites

Natural zeolites as described by Gottardi and Galli

1. Natrolite group (Zeolites with 4=1 building units)
- 1.1.1 Natrolite
  - Para-natrolite
  - Tetra-natrolite
- 1.1.2 Mesolite
- 1.1.3 Scolecite
- 1.1.4 Gonnardite
- 1.2.1 Edingtonite
- 1.3.1 Thomsonite

2. Analcime group
- 2.1.1 Analcime
  - 2.1.2 Wairakite
  - 2.1.3 Hsiang-hualite
  - 2.1.4 Viséite
- 2.2.1 Laumontite
  - Leonhardite
- 2.3.1 Roggianite
- 2.4.1 Yugawaralite
- 2.5.1 Parthéite*

3. Zeolites with double connected four rings
- 3.1.1 Gismondine
  - 3.1.2 Garronite
  - 3.1.3 Amicite
  - 3.1.4 Gobbinsite
- 3.2.1 Phillipsite
  - 3.2.2 Harmotome
- 3.3.1 Merlinoite
- 3.4.1 Mazzite
- 3.5.1 Paulingite

4. Zeolites with six ring building units
- 4.1.1 Gmelinite
- 4.2.1 Chabazite
  - 4.2.2 Will-hendersonite
- 4.3.1 Levyne
- 4.4.1 Erionite
- 4.5.1 Offretite
- 4.6.1 Faujasite
- 4.7.1 Goosecreekite

5. Zeolites with Mordenite framework (5-1 building units)
- 5.1.1 Mordenite
- 5.2.1 Dachiardite
- 5.3.1 Epistilbite
- 5.4.1 Ferrierite
- 5.5.1 Bikitaite

6. Zeolites with Heulandite frameworks (4-4=1 building units)
- 6.1.1 Heulandite
  - 6.1.2 Clinoptilolite
- 6.2.1 Stilbite
  - 6.2.2 Stellerite
  - 6.2.3 Barrerite
- 6.3.1 Brewsterite

7. Zeolites with unknown structure types
- 7.1.1 Cowlesite

* not synthesized

Natural zeolites described by Gottardi and Galli[1] are grouped in first line according to the old mineralogical morphological-structural system which has now been fully adapted to the secondary building unit (SBU)[5] classification. The SBUs are build up by spokes which symbolize the connection between the centers of two $(Si,Al)O_4$ tetrahedrons. An integral number of SBUs is present in a unit cell, but in some cases a given zeolite framework type can be represented by more than one SBU type. The concept is used in order to clarify the complex framework topology of these types of tecto- silicates to which alumino-silicate zeolites belong. They are only theoretical topological building units and should not be considered to be or equated with species that may be in the solution/gel during the crystallization of a zeolitic material[5]. The majority of todays known zeolites (natural and synthetic) can be grouped according to the following eighteen SBU types of figure 2.1.

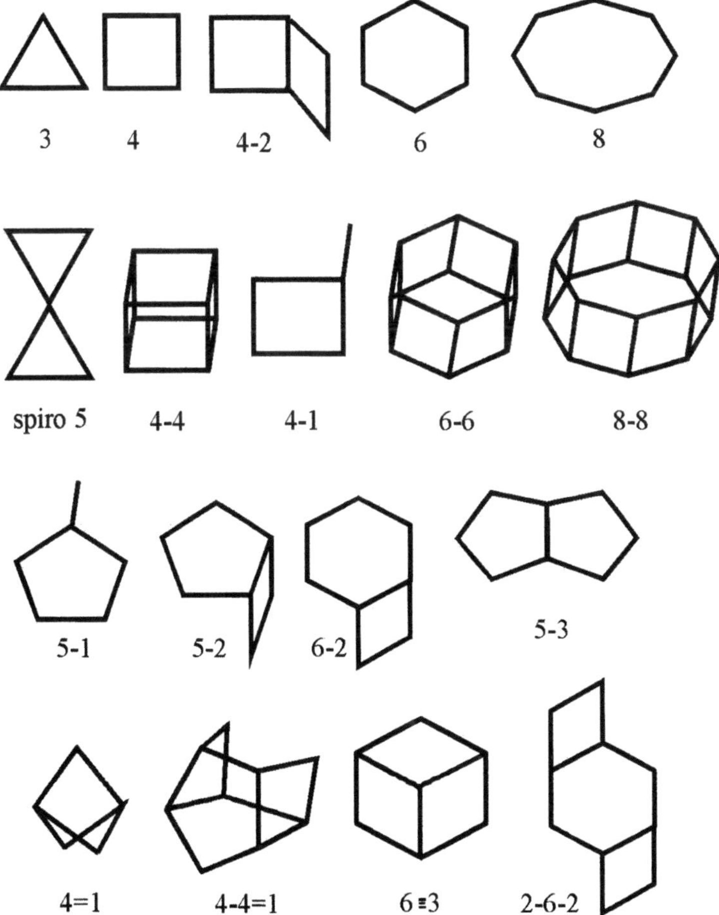

Figure 2.1: The secondary building units and their symbols (reproduced in parts from ATLAS OF ZEOLITE FRAMEWORK TYPES [5], with kind permission).

2.2 Chemical Classification of the Zeolites

A large variety of chemical compositions can be found in natural zeolites. Although the Al/Si ratio in the alumino-silicate framework is quite variable, it is rather restricted compared to zeolite phases obtained by different low pressure hydrothermal synthesis methods[2,9,10,11,5,12]. In nature, the alumino-silicate framework might be doped by transition metal cations in tetrahedral positions. In some cases they play a decisive role during the synthesis in order to obtain the desired zeolite phase. However, their role as a kind of structure directing cation is not yet clarified and still a field of ongoing research. Alkali metal- and alkaline earth metals are essentially the charge compensating cations of the void system in nature. Their kind and concentration are crucial during the synthesis process in order to obtain the desired zeolite phase, although stoichiometry and doping of the framework ions are as important.

Chemical compositions of natural zeolite phases shown in table 2.1-2.3 have been compiled from references[1,5] and [13] applying the actual IUPAC recommendations[4,14]. Data on a respective zeolite stoichiometry in detail can directly be taken from the references given in chapter 5.

TABLE 2.1 The chemical composition of natural -and synthesized natural-alumino-silicate based zeolites: alkali cation zeolites

Zeolite code	name	SBUs	Al:Si ratio[+]	idealized chemical formula	z[*]
NAT	Li-Natrolite	4=1	1:1.5	$|Li^+_2 (H_2O)_2|\ [Al_2Si_3O_{10}]$	1
CAN	Cancrinite	4, 6	1:1	$|Na^+_8 (H_2O)_{2.66}|\ [Al_6Si_6O_{24} (OH)_2]$	1
SOD	Hydroxy-Sodalite	4, 6	1:1	$|Na^+_8 (H_2O)_2|\ [Al_6Si_6O_{24} (OH)_2]$	1
EDI	Zeolite F	4=1	1:1	$|Na^+_5 (H_2O)_9|\ [Al_5Si_5O_{20}]$	1
NAT	Natrolite	4=1	1:1.5	$|Na^+_{16} (H_2O)_{16}|\ [Al_{16}Si_{24}O_{80}]$	1
GIS	Gobbinsite	4-ring	1:2.2	$|Na^+_5 (H_2O)_{11}|\ [Al_5Si_{11}O_{32}]$	1
GME	Gmelinite	6, 6-6	1:2	$|Na^+_2 (H_2O)_5|\ [Al_2Si_4O_{12}]$	4
ANA	Analcime	4, 6	1:2	$|Na^+_{16} (H_2O)_{16}|\ [Al_{16}Si_{32}O_{96}]$	1
CHA	Na-Chabazite	6-ring	1:2.22	$|Na^+_{3.72} (H_2O)_{9.7}|\ [Al_{3.72}Si_{8.28}O_{24}]$	1
FAU	Na-zeolite Y	6-6	1:2.76	$|Na^+_{51} (H_2O)_{7.83}|\ [Al_{51}Si_{141}O_{348}]$	1
ERI	Erionite	4, 6	1:3	$|Na^+_9 (H_2O)_{27}|\ [Al_9Si_{27}O_{72}]$	1
HEU	Na-heulandite	4-4=1	1:3.04	$|Na^+_{8.9} (H_2O)_{26}|\ [Al_{8.9}Si_{27.1}O_{72}]$	1
STI	Barrerite	4-4=1	1:3.5	$|Na^+_8 (H_2O)_{26}|\ [Al_8Si_{28}O_{72}]$	2
VSV	Gaultite	combin.	1:3.5	$|Na^+_4 (H_2O)_5|\ [Zn_2Si_7O_{18}]$	8
EDI	Zeolite K-F	4=1	1:1	$|K^+_{13} (H_2O)_{13}|\ [Al_{10}Si_{10}O_{40} (OH)_3]$	2
MON	Montesommaite	4	1 :2.55	$|K^+_9 (H_2O)_{10}|\ [Al_9Si_{23}O_{64}]$	1
HEU	K-heulandite	4-4=1	1:3	$|K^+_9 (H_2O)_{18}|\ [Al_9Si_{27}O_{72}]$	1
DAC	K-dachiardite	5-1	1:3.8	$|K^+_5 (H_2O)_{12}|\ [Al_5Si_{19}O_{48}]$	1

[+] (Al^{3+},Zn^{2+}) as Al ;   [*] = number of formula units per unit cell

TABLE 2.2 The chemical composition of natural -and synthesized natural-alumino-silicate based zeolites: alkaline earth cation zeolites

Zeolite code	name	SBUs	Al:Si ratio	idealized chemical formula	z^*
SOD	Bicculite	4, 6	1:0.5	$|Ca^{2+}|\,[Al_2SiO_6\,(OH)_2]$	4
-RON	Roggianite	4	1:2	$|Ca^{2+}_{16}\,(H_2O)_{16}|\,[Al_{16}Si_{32}O_{88}\,(OH)_{16}]$	1
PAR	Parthèite	4	1:1	$|Ca^{2+}_2\,(H_2O)_4|\,[Al_4Si_4O_{15}\,(OH)_2]$	4
GIS	Gismondine	4	1:1	$|Ca^{2+}\,(H_2O)_4|\,[Al_2Si_2O_8]$	4
LIO	Liottite	6	1:1	$|Ca^{2+}_9\,(H_2O)_2|\,[Al_{18}Si_{18}O_{72}]$	1
NAT	Scolecite	4=1	1:1.5	$|Ca^{2+}\,(H_2O)_3|\,[Al_2Si_3O_{10}]$	4
GIS	Ca-garronite	4	1:1.66	$|Ca^{2+}_3\,(H_2O)_{12.5}|\,[Al_6Si_{10}O_{32}]$	1
FAU	Ca-zeolite X	6-6	1:1.33	$|Ca^{2+}_{40}\,(H_2O)_4|\,[Al_{80}Si_{112}O_{384}]$	1
ANA	Wairakite	4, 6	1:2	$|Ca^{2+}_8\,(H_2O)_{16}|\,[Al_{16}Si_{32}O_{96}]$	1
LAU	Laumontite	4, 6	1:2	$|Ca^{2+}\,(H_2O)_3|\,[Al_2Si_4O_{12}]$	4
CHA	Chabazite	6	1:2	$|Ca^{2+}_2\,(H_2O)_{13}|\,[Al_4Si_8O_{24}]$	1
LEV	Levyne	6	1:2	$|Ca^{2+}_9\,(H_2O)_{50}|\,[Al_{18}Si_{36}O_{108}]$	3
PHI	Ca-harmotome	4	1:2	$|Ca^{2+}_9\,(H_2O)_{50}|\,[Al_{18}Si_{36}O_{108}]$	1
STI	Stellerite	4-4=1	1:2.2	$|Ca^{2+}_{2.5}\,(H_2O)_{12}|\,[Al_5Si_{11}O_{32}]$	1
YUG	Yugawaralite	4, 8	1:3	$|Ca^{2+}\,(H_2O)_4|\,[Al_2Si_6O_{16}]$	2
GOO	Goosecreekite	6-2	1:3	$|Ca^{2+}\,(H_2O)_5|\,[Al_2Si_6O_{16}]$	2
EPI	Epistilbite	5-1	1:3	$|Ca^{2+}_3\,(H_2O)_{15}|\,[Al_6Si_{18}O_{48}]$	1
***BEA**	Tschernichite	combin.	1:3	$|Ca^{2+}\,(H_2O)_8|\,[Al_2Si_6O_{16}]$	8
HEU	Heulandite	4-4=1	1:3.5	$|Ca^{2+}\,(H_2O)_6|\,[Al_2Si_7O_{18}]$	4
MOR	Ca-mordenite	5-1	1:5	$|Ca^{2+}_4\,(H_2O)_{31}|\,[Al_8Si_{40}O_{96}]$	1
BRE	Ba-brewsterite	4	1:3	$|Ba^{2+}\,(H_2O)_5|\,[Al_2Si_6O_{16}]$	2
EDI	Edingtonite	4=1	1:1.5	$|Ba^{2+}_2\,(H_2O)_8|\,[Al_4Si_6O_{20}]$	1
GME	Ba-Gmelinite	6	1:2	$|Ba^{2+}_4\,(H_2O)_{19.12}|\,[Al_8Si_{16}O_{48}]$	1
CHA	Ba-Chabazite	6	1:2.22	$|Ba^{2+}_{1.86}\,(H_2O)_{7.4}|\,[Al_{3.72}Si_{8.28}O_{24}]$	1
PHI	Harmotome	4	1:3	$|Ba^{2+}_2\,(H_2O)_{12}|\,[Al_4Si_{12}O_{32}]$	1
BRE	Sr-brewsterite	4	1:3	$|Sr^{2+}\,(H_2O)_5|\,[Al_2Si_6O_{16}]$	2

* = number of formula units per unit cell

TABLE 2.3 a) The chemical composition of natural –and synthesized natural-alumino-silicate based zeolites: mixed cation zeolites

zeolite code	name	SBUs	Al:Si ratio[+]	idealized chemical formula	z[*]
ANA	Hsianghualite	4, 6	1:1	$\mid Li^+_{16} Ca^{2+}_{24}\mid [Be_{24}Si_{24}O_{96}(F)_6]$	1
ANA	Viséite	4, 6	1:0.6	$\mid Na^+_2 Ca^{2+}_{10}(H_2O)_{16}\mid [Al_{20}Si_6P_{10}O_{60}(OH)_{36}]$	1
THO	Thomsonite	4=1	1:1	$\mid Na^+_2 Ca^{2+}_4 (H_2O)_{12}\mid [Al_{10}Si_{10}O_{40}]$	2
GIS	Amicite	4	1:1	$\mid Na^+_4 K^+_4 (H_2O)_{10}\mid [Al_8Si_8O_{32}]$	1
CHA	Will-hendersonite	6	1:1	$\mid K^+_2 Ca^{2+}_2 (H_2O)_{10}\mid [Al_6Si_6O_{24}]$	1
TSC	Tschörnerite	4, 6, 6-6	1:1	$\mid Ca^{2+}_{64} (K^+_2, Ca^{2+}, Sr^{2+}, Ba^{2+})_{48} Cu^{2+}_{48} (OH^-)_{128} (H_2O)_x\mid [Al_{24}Si_{24}O_{96}]$	16
FRA	Franzinite	6	1:1	$\mid Ca^{2+}_{10} (K^+,Na^+)_{30} (SO_4)^{2-}_{10} (H_2O)_2\mid [Al_{30}Si_{30}O_{120}]$	1
AFG	Afghanite	4, 6	1:1	$\mid Ca^{2+}_{10} Na^+_2 (Cl^-)_2 (SO_4)^{2-}_5 (H_2O)_4\mid [Al_{24}Si_{24}O_{96}]$	1
EAB	Bellbergite	4, 6	1:1	$\mid Na^+ Ba^{2+} Sr^{2+})_2 Sr^{2+}_2 Ca^{2+}_2 (Na^+ Ca^{2+})_4 (H_2O)_{30}\mid [Al_{18}Si_{18}O_{72}]$	1
NAT	Gonnardite	4=1	1:1.22	$\mid Na^+_5 Ca^{2+}_2 (H_2O)_{10}\mid [Al_9Si_{11}O_{40}]$	1
-WEN	Wenkite	combin.	1:1.5	$\mid Ba^{2+}_4 (Ca^{2+},Na^+_2)_3 (SO_4)^{2-}_3 (H_2O)\mid [Al_8Si_{12}O_{39}(OH)_2]$	1
NAT	Mesolite	4=1	1:1.5	$\mid Na^+_{16} Ca^{2+}_{16} (H_2O)_{64}\mid [Al_{48}Si_{72}O_{240}]$	1
GIS	Garronite	4	1:1.66	$\mid Na^+ Ca^{2+}_{2.5} (H_2O)_{13}\mid [Al_6Si_{10}O_{32}]$	1
PHI	Phillipsite	4	1:1.66	$\mid K^+_2 (Ca^{2+}_{0.5}, Na^+)_4 (H_2O)_{12}\mid [Al_6Si_{10}O_{32}]$	2
LEV	Levyne	6	1:2	$\mid Na^+ Ca^{2+}_{2.5} (H_2O)_{18}\mid [Al_6Si_{12}O_{36}]$	3
LTL	Perialite	6, 8	1:2	$\mid Na^+ K^+_9 (Ca^{2+}, Sr^{2+}) (H_2O)_{15}\mid [Al_{12}Si_{24}O_{72}]$	1
PHI	Harmotome	4	1:2.2	$\mid Ba^{2+}_2 (Ca^{2+}_{0.5}, Na^+) (H_2O)_{12}\mid [Al_5Si_{11}O_{32}]$	1
FAU	Faujasite	6-6	1:2.2	$\mid Na^+_{20} Ca^{2+}_{12} Mg^{2+}_8 (H_2O)_{235}\mid [Al_{60}Si_{132}O_{384}]$	1
-CHI	Chiavenite	5-2	1:2.5	$\mid Ca^{2+}_4 Mn^{2+}_4 (H_2O)_2\mid [Be_2Si_5O_{13}(OH)_2]$	4
MER	Merlinoite	4, 8-8	1:2.55	$\mid (K^+,Na^+)_5 (Ba^{2+},Ca^{2+})_2 (H_2O)_{24}\mid [Al_9Si_{23}O_{64}]$	1
MAZ	Mazzite	4, 5-1	1:2.6	$\mid K^+_3 Ca^{2+}_{1.5} Mg^{2+}_2 (H_2O)_{28}\mid [Al_{10}Si_{26}O_{72}]$	1
OFF	Offretite	6	1:2.6	$\mid K^+ Ca^{2+} Mg^{2+} (H_2O)_{15}\mid [Al_5Si_{13}O_{36}]$	1
HEU	Heulandite	4-4=1	1:3	$\mid (Na^+,K^+) Ca^{2+}_4 (H_2O)_{24}\mid [Al_9Si_{27}O_{72}]$	1
STI	Stilbite	4-4=1	1:3	$\mid Na^+ Ca^{2+}_4 (H_2O)_{30}\mid [Al_9Si_{27}O_{72}]$	1
PAU	Paulingite	4	1:3.08	$\mid Na^+_{12} K^+_{68} Ca^{2+}_{41} (H_2O)_{705}\mid [Al_{162}Si_{500}O_{1344}]$	1
LOV	Lovdarite	combin.	1:3.5	$\mid Na^+_4 K^+_{12} (H_2O)_{18}\mid [Be_8Si_{28}O_{72}]$	1
ERI	Erionite	4, 6	1:3.5	$\mid Na^+ K^+_2 Mg^{2+} Ca^{2+}_{1.5} (H_2O)_{28}\mid [Al_8Si_{28}O_{72}]$	1
BOG	Boggsite	4, 6	1:4.33	$\mid Na^+_4 Ca^{2+}_7 (H_2O)_{74}\mid [Al_{18}Si_{78}O_{192}]$	1

[+] (Al^{3+},Be^{2+}) as Al, (P^{5+},Si^{4+}) as Si;   [*] = number of formula units per unit cell

TABLE 2.3 b) The chemical composition of natural –and synthesized natural-alumino-silicate based zeolites: mixed cation zeolites

zeolite code	name	SBUs	Al:Si ratio[+]	idealized chemical formula	z[*]
MOR	Mordenite	5-1	1:5	$\mid Na^+_3 K^+ Ca^{2+}_2 (H_2O)_{28} \mid [Al_8 Si_{40} O_{96}]$	1
DAC	Dachiardite	5-1	1:5	$\mid (Na^+, K^+, Ca^{2+}_{0.5})_4 (H_2O)_{18} \mid [Al_4 Si_{20} O_{48}]$	1
HEU	Clinoptilolite	4-4=1	1:5	$\mid (Na^+, K^+)_6 (H_2O)_{20} \mid [Al_6 Si_{30} O_{72}]$	1
FER	Ferrierite	5-1	1:5	$\mid (Na,K)^+ Mg^{2+}_2 Ca^{2+}_{0.5} (H_2O)_{20} \mid [Al_6 Si_{30} O_{72}]$	1
TER	Terranovaite	2-6-2	1:5.5	$\mid Na^+_{4.2} K^+_{0.25} Ca^{2+}_{3.7} Mg^{2+}_{0.2} (H_2O)_{29} \mid$ $[Al_{12.3} Si_{67.7} O_{160}]$	1
NES	Gottardiite	5-3	1:6.15	$\mid Na^+_3 Mg^{2+}_3 Ca^{2+}_5 (H_2O)_{93} \mid [Al_{19} Si_{117} O_{272}]$	1
MFI	Mutinaite	5-1	1:7.58	$\mid Na^+_3 Ca^{2+}_{3.78} Mg^{2+}_{0.21} (H_2O)_{60} \mid [Al_{11.2} Si_{84.9} O_{192}]$	1

[+] (Al^{3+}, Be^{2+}) as Al, (P^{5+}, Si^{4+}) as Si;   [*] = number of formula units per unit cell

Pure phosphates such as the « AlPOs » and the « BePOs » have been omitted in this overview on the chemical composition of alumino-silicate based zeolites. At the time when Gottardi and Galli published NATURAL ZEOLITES they were not respected as proper zeolite phases in terms of a zeolite definition.

It is also interesting to know that most of the natural zeolites occur as mixed cation zeolites. Our zeolite reconstruction work has shown that the presence of several different charge compensating cations is one of the essential factors for the templation of the zeolite void system - during the natural formation as well as during the synthesis in the laboratory. It has also shown that the chemical compositions of the natural minerals correspond almost perfectly to the chemical composition during the formation: The zeolite reconstruction work gave the proof that in most cases no secondary ion exchange after the natural zeolite formation has been taken place.

2.3 Secondary Building Unit Classification

Among the secondary building units existing (shown in figure 2.1) few characteristic ones classify the natural zeolites described by Gottardi and Galli[1]:

Zeolites with 4=1 Chains

Zeolites with Singly Connected 4-Ring Chains

Zeolites with Double Connected 4-Ring Chains

Zeolites with 6-Rings

Zeolites with 5-1 Building Units

Zeolites with 4-4=1 Building Units

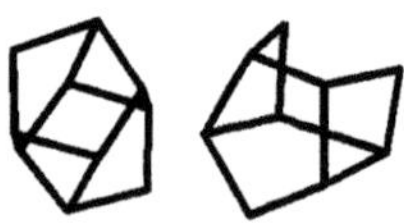

Goosecreekite (GOO, 6-2 SBUs), Partheite (-PAR, 4 SBUs) and Cowlesite were the three zeolites with unknown structure types in 1986, the publication year of NATURAL ZEOLITES[1]. Today, only Cowlesite remains unclassified due to problems regarding the resolution of the structure as only small quantities of polycrystalline natural samples are available while the synthesis always gives phase mixtures not appropriate for structural refinement.

2.4 Channel, Cage and Cavity Systems

The outstanding property of zeolites is the presence of a structure inherent void system giving them their specific unique « zeolitic » properties.

Channel systems in zeolitic materials reach different levels of complexity.

Channel systems are always passing along crystallographic main directions which are directly related to the morphology of idiomorphous crystals.

One dimensional channel systems are not interconnected and pass along a given crystallographic direction as shown in figure 2.2.

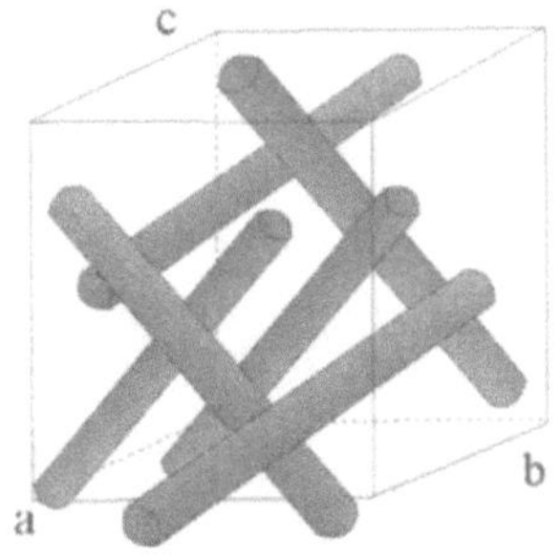

Figure 2.2: The « 1d » channel system of Analcime -type zeolites (ANA).

Two dimensional channel systems are passing along two crystallographic main directions (see figure 2.3). For geometric reasons at the channels interconnection additional voids (cavities) with bigger space than the original channels are present. In some cases the two dimensional channel system can weaken the framework stability in this preferential direction. This was the reason why the term « leafy zeolites » for these kinds of framework types of natural zeolites was used (in terms of their visual morphology) to which e.g. Stilbite belongs.

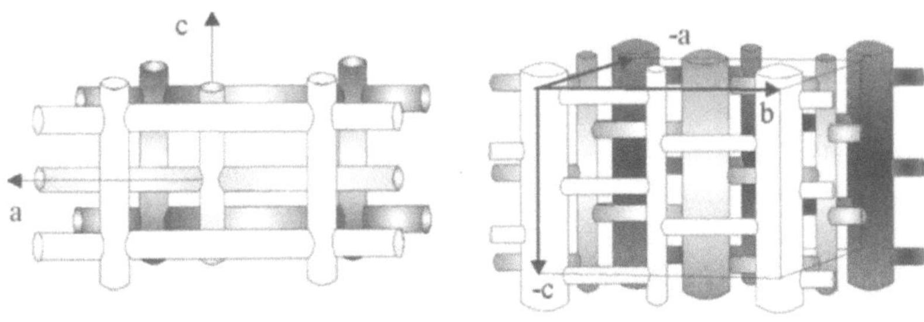

Figure 2.3: The « 2d » channel systems of Stilbite (STI, left) and Mordenite (MOR, right).

Three dimensional channel systems might be interconnected in different ways as can be seen in the examples of figure 2.4. Cavities bigger than the original channel size are present at the interconnection points, but they can also be part of the channel system as for example in the case of zeolite Chabazite. Zeolite Paulingite (fig. 2.5) is an example for a three dimensional channel system set into another without any interconnection.

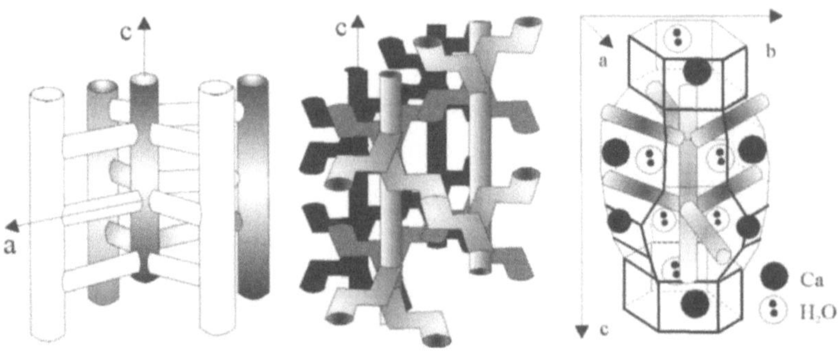

Figure 2.4: The « 3d » channel systems of Brewsterite (BRE, left), Natrolite (NAT, middle) and Chabazite (CHA, right) represented in different modes.

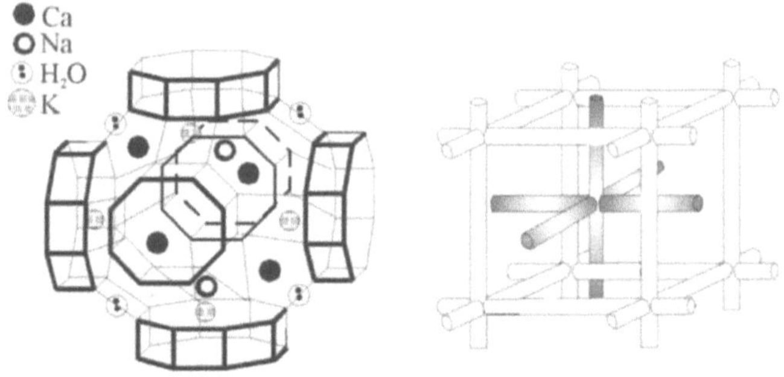

Figure 2.5: The « 3d » channel system of Paulingite (PAU) in two different modes of representation.

Besides channel systems passing the zeolite structure straight or in an intersected way, interconnected cavities might also constitute channel systems. Although the effective channel diameter is in this case limited by the diameter of the interconnective opening, the cavities might serve as microreactors, where organic molecules small enough to access enter and react inside. Only products can leave which are correspondingly small enough to pass the opening.

Figure 2.6 shows the Faujasite structure in three different modes revealing the channel like cavity-cage interconnectivity.

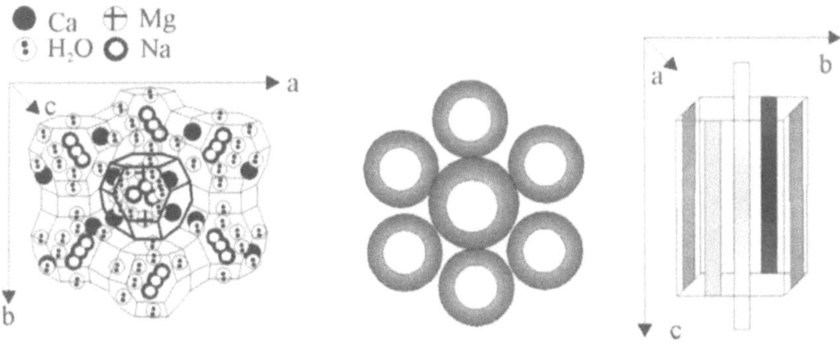

Figure 2.6: The « 3d » interconnected cavity system of Faujasite (FAU) : left structural view with Sodalite cages surrounding the FAU-supercavity ; middle: arrangement of the interconnected FAU supercavities ; right : channels in c-direction formed by cage-cavity arrangements.

2.4.1 Channel System Classification

The notation used in ATLAS OF ZEOLITE FRAMEWORK TYPES[5] for channel system classification has been directly employed, where each system of equivalent channels is described by
• the channel direction (relative to the axes of the type material structure),
• the number of tetrahedrally coordinated atoms (in bold type) forming the rings controlling diffusion through the channels, and
• the crystallographic free diameters of the channels (here: in Angstroem units).
The number of asterisks in the notation indicates whether the channel system is one-, two- or three-dimensional.
In most cases, the smaller rings simply form openings which connect larger cavities. Interconnecting channel systems are separated by a double arrow (↔). In contrast, a vertical bar (|) means that there is no direct access from one channel system to the other. <100> means there are channels parallel to all crystallographically equivalent directions of the cubic structure, i.e., along x, y and z. Further details of the notation can be taken from the ATLAS OF ZEOLITE FRAMEWORK TYPES [5], while an accurate, extensive but rather complicated description of the pore system can be found in[4,14].

TABLE 2.4 The classification of natural and synthesized natural zeolites according to the size of the channel system (data from[5]).

Code	Zeolite phase	Channel system in first -	Second and Third crystallographic direction
FAU	Faujasite	<111> **12** 7.4***	
MAZ	Mazzite	[001] **12** 7.4* |	[001] **8** 3.4 x 5.6*
LTL	Perlialite	[001] **12** 7.1*	
BOG	Boggsite	[100] **12** 7.0 x 7.0* ↔	[010] **10** 5.5 x 5.8*
GME	Gmelinite	[0001] **12** 7.0* ↔	⊥[0001] **8** 3.6 x 3.9**
OFF	Offretite	[0001] **12** 6.7* ↔	⊥[0001] **8** 3.6 x 4.9**
MOR	Mordenite	[001] **12** 6.5 x 7.0* ↔	[010] **8** 2.6 x 5.7*
*BEA	Tschernichite	[001] **12** 5.6 x 5.6* ↔	<100> **12** 6.6 x 6.7**
CAN	Cancrinite	[001] **12** 5.9 x 5.9*	
MFI	Mutinaite	{[010] **10** 5.3 x 5.6 ↔	[100] **10** 5.1 x 5.5}***
TER	Terranovaite	[100] **10** 5.0 x 5.5* ↔	[001] **10** 4.2 x 7.0*
STI	Stilbite	[100] **10** 4.9 x 6.1 ↔	[101] **8** 2.7 x 5.6*
NES	Gottardiite	[100] **10** 4.7 x 6.0**	
FER	Ferrierite	[001] **10** 4.2 x 5.4* ↔	[010] **8** 3.5 x 4.8*
TSC	Tschörnerite	<100> **8** 4.2 x 4.2** ↔	<110> **8** 3.1 x 5.6***
-RON	Roggianite	[001] **12** 4.2*	
LAU	Laumontite	[100] **10** 4.0 x 5.3*	
-CHI	Chiavennite	[001] **9** 3.9 x 4.3*	
CHA	Chabazite	⊥[0001] **8** 3.8 x 3.8***	
PAU	Paulingite	<100> **8** 3.8*** |	<100> **8** 3.8***
EPI	Epistilbite	[100] **10** 3.4 x 5.6*↔	[001] **8** 3.7 x 5.2*
EAB	Bellbergite	⊥[001] **8** 3.7 x 5.1**	
DAC	Dachiardite	[010] **10** 3.4 x 5.3* ↔	[001] **8** 3.7 x 4.8*
VSV	Gaultite	[011] **9** 3.3 x 4.5* ↔	[01$\bar{1}$] **9** 3.3 x 4.5* ↔ [10$\bar{1}$] **8** 3.7 x 3.7*
ERI	Erionite	⊥[0001] **8** 3.6 x 5.1***	
LEV	Levyne	⊥[001] **8** 3.6 x 4.8**	
WEI	Weinebergite	[001] **10** 3.1 x 5.4* ↔	[100] **8** 3.3 x 5.0*
LOV	Lovdarite	[010] **9** 3.2 x 4.4* ↔	[001] **9** 3.2 x 3.7* ↔ [100] **8** 3.6 x 3.7*
PHI	Phillipsite	[100] **8** 3.6* ↔	[010] **8** 3.0 x 4.3* ↔ [001] **8** 3.2 x 3.3*
RHO	Pahasapaite	<100> **8** 3.6*** |	<100> **8** 3.6***
MON	Montesommaite	[001] **8** 3.6 x 3.6 * ↔	[100] **8** 3.2 x 4.4*
-PAR	Partheite	[001] **10** 3.5 x 6.9*	
MER	Merlinoite	[100] **8** 3.1 x 3.5* ↔	[010] **8** 2.7 x 3.6* ↔ [001]{**8** 3.4 x 5.1* + **8** 3.3 x 3.3*}
HEU	Heulandite	{[001] **10** 3.0 x 7.6* + **8** 3.3 x 4.6*} ↔	[100] **8** 2.6 x 4.7*
YUG	Yugawaralite	[100] **8** 2.8 x 3.6* ↔	[001] **8** 3.1 x 5.0*
GIS	Gismondine	{[100] **8** 3.1 x 4.5* ↔	[010] **8** 2.8 x 4.8}*
BRE	Brewsterite	[100] **8** 2.3 x 5.0* ↔	[001] **8** 2.8 x 4.1*
GOO	Goosecreekite	[100] **8** 2.8 x 4.0* ↔	[010] **8** 2.7 x 4.1* ↔ [001] **8** 2.9 x 4.7*
EDI	Edingtonite	[110] **8** 2.8 x 3.8** ↔	[001] **8** variable*
BIK	Bikitaite	[001] **8** 2.8 x 3.7*	
-WEN	Wenkite	<100> **10** 2.6 x 4.9** ↔	[001] **8** 2.2 x 2.7*
NAT	Natrolite	<100> **8** 2.6 x 3.9** ↔	[001] **8** variable*
THO	Thomsonite	[101] **8** 2.3 x 3.9* ↔	[010] **8** 2.2 x 4.0* ↔ [001] **8** variable
ANA	Analcime	[110] **8** 1.6 x 4.2*** irregular	
SOD	Sodalite	apertures formed by six-rings only	
LIO	Liottite	apertures formed by six-rings only	
FRA	Franzinite	apertures formed by six-rings only	
AFG	Afghanite	apertures formed by six-rings only	

2.4.2 Zeolite Cages and Cavities

The IUPAC recommendations[4,14] define in terms of the zeolite notation n-rings forming faces of a polyhedral pore as windows, while cages are voids consisting of polyhedral units too narrow to be penetrated by molecules larger than water molecules. Cavities in these terms are polyhedral pores (voids) having at least one face defined by a window to be penetrated by molecules bigger than water molecules, but which are not forming a channel (meaning: do not pass the structure from one end to the other). Figure 2.7 gives a selection of windows, cages and cavities.

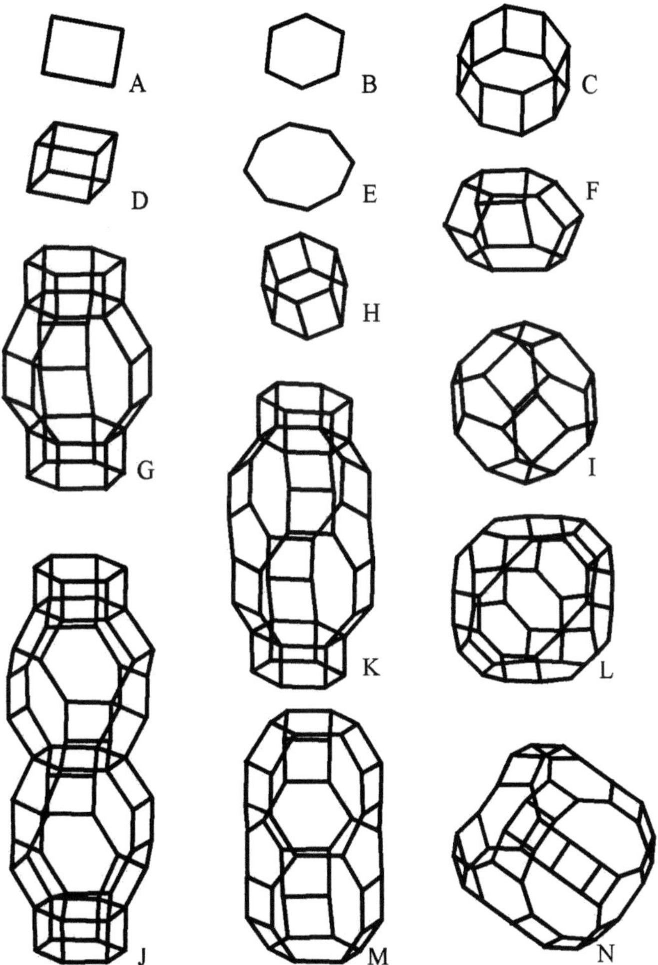

Figure 2.7: Selected zeolite windows, cages and cavities: A = 4-ring (window opening to cage, cavity or channel) , B = 6-ring (window opening to cage, cavity or channel), C = 8-8-cavity (δ-cavity), D = 4-4-cage, E = 8-ring (window opening to cage, cavity or channel), F = Cancrinite cage (ε-cage), G = Gmelinite cavity (γ-cavity), H = 6-6-cavity, I = Sodalite cage (β-cage), J = Levyne cavity, K = Chabazite cavity, L = α-cavity, M = Erionite cavity, N = Faujasite supercavity.

2.5 Zeolitic Water Related to Cations

Charge compensating cations in extraframework sites of the zeolite can be found in the following positions[15]:

- co-ordinated by framework oxygen
- co-ordination by framework oxygen at nearly opposite sides
- co-ordination by framework oxygen on the one hand side and water molecules on the other (partially hydrated at one side within the zeolitic void)
- complete co-ordination of the cation by water molecules (fully hydrated within the zeolitic void)

A compilation of extra framework sites can be found in[16]. The water and extraframework positions might also be taken from the positions in the asymetric unit (see e.g. ICSD data base data[13]).

In the activated state, under conditions where the zeolites are mostly applied for catalytic purposes, they are almost de-hydrated. Cations which were originally surrounded by a hydration sphere within the void system « stick » to the channel or cavity wall. The original zeolite properties might, therefore, be largely -and in some cases even irreversibly- disturbed.

3. HYDROTHERMAL SYNTHESIS UNDER PRESSURE

3.1 Physical Chemistry of Hydrothermal Synthesis Under Pressure

The hydrothermal synthesis method under high pressure is able to simulate the natural synthesis conditions and to "reconstruct" natural zeolites, but also to provide a way to obtain substituted zeolites, e.g. by transition metal ions, where only very few cases are known in nature (e.g. zeolite VSV-Gaultite). This chapter is intended to describe the elementary physical chemistry of the synthesis process, i.e. thermodynamics and kinetics of precursor dissolution and of zeolite nucleation and growth. Water is the solvent universally used in the zeolite reconstruction work. Temperature, pressure and reaction time are the three principal physical parameters in hydrothermal processing.
Given that the synthesis is performed in closed high pressure autoclaves, a major problem for the experimental determination of the synthesis mechanism is the difficulty of in situ measurements, indicating which elementary species are present, how they evolve etc. In the following the essential knowledge on dissolution/crystallization processes is summarized and examined how high pressure hydrothermal conditions might affect them.

3.1.1 Autogeneous and High Pressure Hydrothermal Conditions

The phase diagram of pure water is represented in figure 3.1[17]. The autogenous pressure range is fixed by the two-phase equilibrium line between boiling point (1 bar, 100°C) and critical point (221bar, 374°C). In other words, as long as liquid and vapour coexist, the pressure is fixed by the temperature of the experiment. A great number of conventional synthesis experiments are performed in this region. Above the critical point however, only one fluid phase exists and the pressure can be enhanced externally to much higher values, corresponding to so-called supercritical conditions.
Under these circumstances, the solubility of most inorganic substances increases tremendously for both thermodynamic and kinetic reasons[18]. This permits e. g. supercritical extraction by fluid carbon dioxide[19,] or, for the present purpose, dissolution of inorganic substances typically insoluble in water under normal atmospheric conditions, such as alumino-silicate glasses, because the solubility of inorganic materials is generally rising with pressure, as discussed below. This is the fundament of the high pressure synthesis method described here.

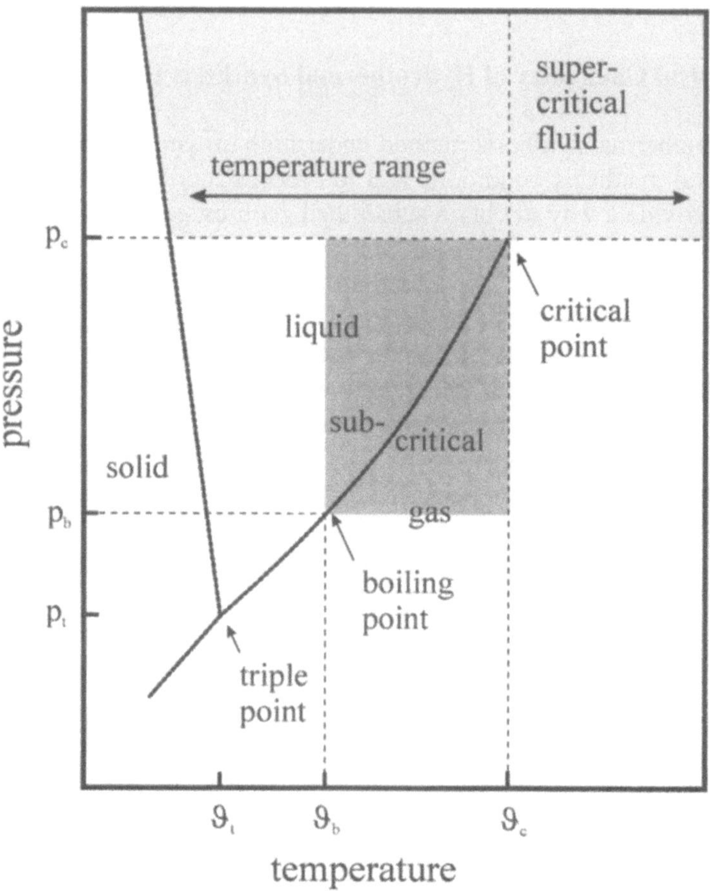

Figure 3.1: The phase diagram of water indicating the pressure and temperature rage of hydrothermal syntheses under very high pressure.

Hydrothermal conditions also modify considerably the aqueous solution chemistry, favoring small complex-structures with lower symmetry, such as partially hydrated central ions, cation-anion associates etc.[20]. The properties of the solvent are strongly influenced by temperature and pressure. The dielectric constant of water decreases with temperature and increases with pressure: hydrothermal solutions present therefore a low dielectric constant and electrolytes tend to form ion pairs. The viscosity of the solvent decreases with temperature leading to larger ion mobilities and enhanced reaction kinetics. Furthermore, temperature enhancement leads to a higher self-dissociation of water with strongly modified acidity/basicity constants, as shown in figure 3.2. The strong enhancement of molecular interactions and reduction of kinetic barriers are probably the most important factors for the amazing results of high pressure hydrothermal synthesis.

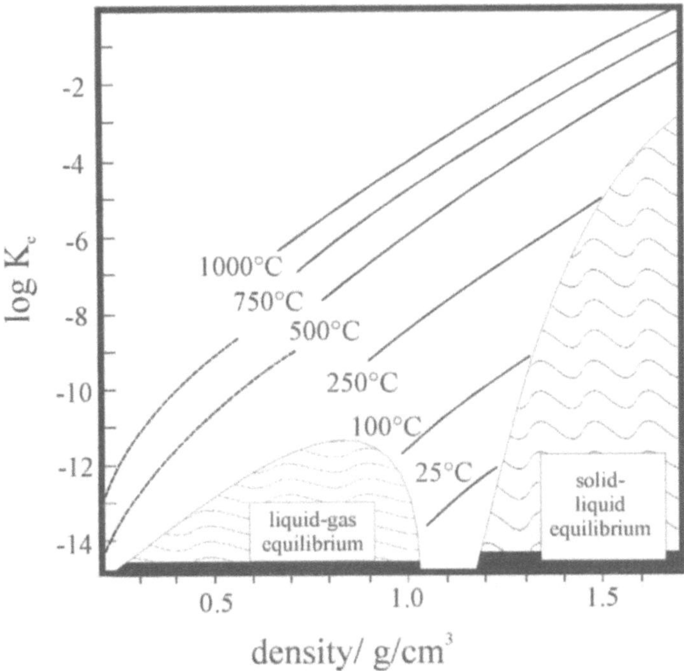

Figure 3.2: The ion product of water dependent on density at different temperatures[20].

3.1.2 Thermodynamics and Kinetics of Dissolution of Iono-covalent Solids

The equilibrium situation (solubility limit) is described by a minimum of the Gibbs free energy of the system water+solute. The Gibbs free energy variation ΔG includes energetic as well as entropic contributions, according to the Gibbs-Helmholtz equation:

$$\Delta G = \Delta H - T\Delta S \tag{1}$$

ΔH is the solution enthalpy and ΔS the solution entropy. The equilibrium solubility of a solute is given by equilibrium between entropic and energetic terms[21].

The entropy term is essentially configurational, resulting from the dispersion of solute in the solvent, due essentially to random thermal motion. This term is always driving the dissolution. The configurational entropy ΔS is relatively easy to quantify, at least in diluted solutions:

$$\Delta S \sim - R \ln x \tag{2}$$

R is the gas constant and x is the molar fraction of solute. In concentrated media, such as those currently observed under hydrothermal conditions however, free ions are not the most probable alternative and ion pairs or higher aggregates must be considered.

The energy term includes several contributions. First, in order for the ions to leave the solid, energy must be spent against the cohesion energy. Second, work must be done against the solvent-solvent interactions, especially the hydrogen bonds in the case of water. Third, energy is gained by the interaction between solvent and dissolved ions, corresponding to an enthalpy of solvatation. In the case of water, this term is called hydration enthalpy. According to classical irreversible thermodynamics, the kinetics of a process can be enhanced by increasing the driving force. The following discussion of thermodynamic factors will also be relevant for the dissolution kinetics.

Lattice or cohesion enthalpy of the solid precursor

For crystalline ionic materials, the lattice energy calculation is relatively easy to perform[22]. Knowing the Madelung constant of a crystal lattice system, the electrostatic energy term, which is derived from Coulomb's law, is about 90% of the total energy and smaller energy contributions, such as repulsive or Van der Waals energy, can often be neglected in first approximation. The charge of the ions forming the lattice is an essential factor: the lattice energy will increase proportional to the product of individual ion charges. If the precursors are amorphous, like in our case, the cohesion energy of glasses can be estimated from solution models and empirical equations, such as those obtained for regular or sub-regular solutions[23].

The previous discussion applies to bulk samples with reduced surface/volume ratio. However, if the precursor solids have a very large surface area, i.e. are present as fine powders or divided solids, surface energy has to be taken into account and the influence of the particle size becomes important for the thermodynamics and kinetics of dissolution. According to the Gibbs-Kelvin equation (see eq. (6)), the equilibrium solubility depends on the particle size: the smaller the particle, the higher its solubility. Furthermore, the surface orientation of crystallites is important, because in crystalline substances the surface energy is anisotropic. In case of glassy precursors however, the generally isotropic properties of glasses reduce the influence of surface effects. Nevertheless, the particle size effect remains.

Hydration enthalpy

The hydration under elevated pressure consists in the formation of aquo complexes of small dimensions where individual ions are fully or partly surrounded by water molecules. Very often, the geometry of aquo-complexes reflects typical units also found in the solid, e.g. octahedral or quadratic environment. The hydration enthalpy depends on the ion radius and charge. Small ions with high charge present in general a large hydration enthalpy. Consequently, the hydration enthalpy decreases inside a column of the periodic table. Transition metal ions have generally a large hydration enthalpy due to strong interactions of d- and f-electrons with water molecules.

In the fluid phase under very high pressure, one can estimate that the hydration shell of ions is reduced in comparison with standard conditions. The incomplete hydration shell induces a high reactivity and ion association phenomena, as discussed previously.

Solution enthalpy

The sum of cohesion energy $\Delta_{coh}H$ and hydration energy $\Delta_{hyd}H$ gives the solution energy $\Delta_{sol}H$.

$$\Delta_{sol}H = \Delta_{coh}H + \Delta_{hyd}H \tag{3}$$

In general, the solubility decreases with increasing ion charge, due to the increased lattice or cohesion energy: oxides are less soluble than halides, salts of alkaline earths are less soluble than alkali salts and so on. Depending on the sign of the solution enthalpy, a temperature increase can enhance the precursor solubility (endothermic dissolution) or reduce it (exothermic dissolution). The higher the solution enthalpy is, the larger is the solubility change with temperature. If glassy and crystalline solids are compared, the latter have higher cohesion energy, due to the perfect crystalline arrangement and smaller inter-ionic distances. Under comparable conditions, the crystalline zeolite is therefore less soluble than the amorphous precursor glass. This is the driving force for zeolite crystallization in our high pressure hydrothermal reconstruction work.

Increase of pressure has been empirically shown to increase the solubility in general. The thermodynamic reason is the generally smaller molar volume and larger compressibility of the fluid compared with the solid. Increasing the pressure will thus favour the presence of fluid phase only, in other words the dissolution of the solid.

3.1.3 Thermodynamics and Kinetics of Nucleation and Growth

Four steps can be distinguished during crystallization from a solution, which is applicable to hydrothermal fluids in principle: i) generation of neutral species in solution, ii) nucleation, iii) growth, iv) ageing.

Formation of neutral species

The first step is the formation of an uncharged species, capable to condense and form a germ of the product. Cation hydroxylation is for example a very simple and normally fast reaction, but the kinetics can be significantly modified if the reaction is made e. g. from complex cations. There is a large body of knowledge on elementary species in "normal" aqueous solutions and some NMR work under high pressure conditions[24,25].

Kinetics of nucleation

The nucleation rate depends on the driving force, i.e. the over-saturation of the solution. This is where the use of amorphous precursors is advantageous, because the glass solubility is higher than that of the crystalline zeolite phase. In other words, an over-saturation versus the crystalline phase can be established during the hydrothermal synthesis process and the nucleation rate can be tailored specifically. This is nicely exemplified in figure 3.3: above a minimum precursor concentration in solution, at about C_{min}, the rate of nucleation increases sharply and nuclei are formed faster the higher the over-saturation. If the dissolution rate is much lower than the nucleation rate (general case in hydrothermal fluids), nucleation leads to a strong decrease of precursor concentration passing a maximum C_{max}. Nucleation stops and the number of particles keeps constant when the precursor concentration falls significantly below the level C_{min} due to further nucleation and simultaneously starting growth.

Classical homogeneous nucleation theory was developed in the 1930's. The basic equations show that the nucleation barrier height is determined by the over-saturation term and a term taking the energy of the product/solution interface into account. The interface energy γ is defined as $\gamma = dG/dA$, where A is the interface area. The chemical potential difference $(\mu_S-\mu_L)$ is a function of the precursor concentration in solution C_L and the product solubility C_S. The ratio C_L/C_S is the oversaturation term $\mathcal{N}$:

$$\Delta G = n\,(\mu_S-\mu_L) + A\gamma = nkT\,\ln(C_S/C_L) + A\gamma = -\,nkT\,\ln\mathcal{N} + A\gamma \qquad (4)$$

n is the number of neutral entities forming the germ (nucleus).

Homogeneous nucleation theory yields the critical germ radius and the activation barrier for nucleation that has to be overcome in order to obtain a stable germ. The critical germ radius is given by the Gibbs-Kelvin type equation:

$$r^* = 2\,\gamma V/(kT\ln\mathcal{N}) \tag{5}$$

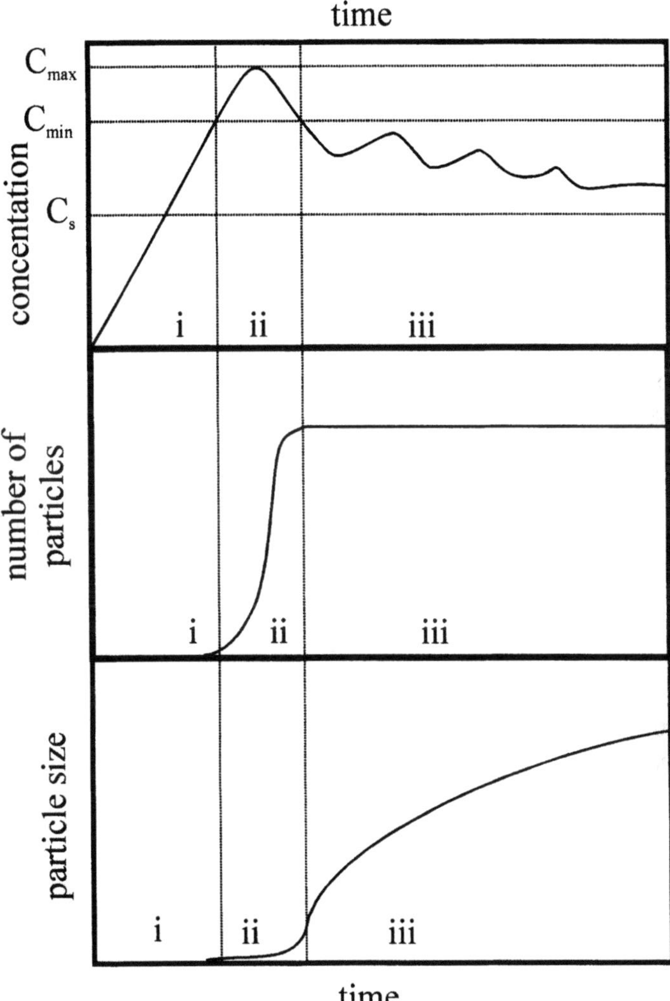

Figure 3.3: Isothermal dissolution-recrystallization scheme for amorphous precursors in hydrothermal aqueous fluids under high pressure: i) generation of neutral species in solution, ii) nucleation, iii) growth (after [20], modified).
Above: concentration dependent on time; C_s=equilibrium solubility; C_{min} = level of minimum over-saturation for sustained nucleation, C_{max} = maximum over-saturation. Middle: number of nuclei evolving to particles in tibe different zones of concentration. Below: Particle size evolution with time: crystal growth essentially in zone iii).

The critical germ radius to create stable nuclei decreases with increasing over-saturation and decreasing interface energy of the nucleating phase. The second reason is the origin of heterogeneous nucleation due to interface energy changes by formation of the nuclei on a substrate. It is often the preferred mechanism: in the case on hand, the zeolite crystallites grow normally on precursor glass substrates. The acceleration of heterogeneous nucleation is related to a reduction of the interface energy term, due to interaction with a substrate, if the interface energy substrate/product is below the interface energy product/solution. Obviously, the choice of a substrate having a low interface energy with the zeolite product can create favourable conditions for heterogeneous nucleation.

Kinetics of growth

At this stage, nuclei growth starts by condensation of neutral species on already formed nuclei, if the concentration in the solution remains below C_{min} as can be seen in figure 3.3. However, nucleation and growth can be simultaneous processes if the precursor concentration remains above C_{min}. The number and size of particles are thus a function of the grade of over-saturation during the complete process. In order to obtain particles with homogeneous size, it is necessary to separate nucleation and growth with only one nucleation step and further regular growth. Fluctuations of concentration below C_{min} have to be taken into account in the experiment, but do not trigger the precipitation of a second crystal generation if C_{min} is not exceeded.

Supersaturation conditions at about C_{min} with simultaneous nucleation and growth can lead to a large size distribution of the particles. In other words, product nucleation must be much faster than precursor dissolution, in order to have a very short nucleation period (general case in hydrothermal processing under high pressure).

The growth rate can be limited by diffusion of species if the incorporation into the germ is very fast, or by other processes, e.g. by a surface reaction preceding incorporation into the germ. Diffusional limitation of growth can be described by equations derived from Fick's first law:[20]

$$dr/dt = D(C - C_S)v/r \tag{6}$$

r is the germ radius, D the diffusion coefficient of the solute, C its concentration and v its molar volume. If surface reactions are rate-determining, the growth equations can be more complicated.[20]

Ageing

Primary particles can age, because they are not necessarily in an equilibrium state, although the long reaction times used in hydrothermal synthesis are favourable for thermodynamically stable phases in principle. The so-called "Oswald ripening" can lead to larger particle size and aggregation.

The ageing process can also lead to changes of morphology or even to phase transitions into thermodynamically rather than kinetically stable phases. A characteristic example taken from the zeolite reconstruction work is shown in figure 3.4. Here, at higher synthesis temperature the thermodynamically stable phase β-eucryptite is observed, while at lower temperature the kinetically favoured phase zeolite ABW is prevailing, both having the same stoichiometry. However, the amount of zeolite ABW decreases continuously with rising synthesis temperature, an indirect proof for the successive conversion of ABW to β-eucryptite by ageing.

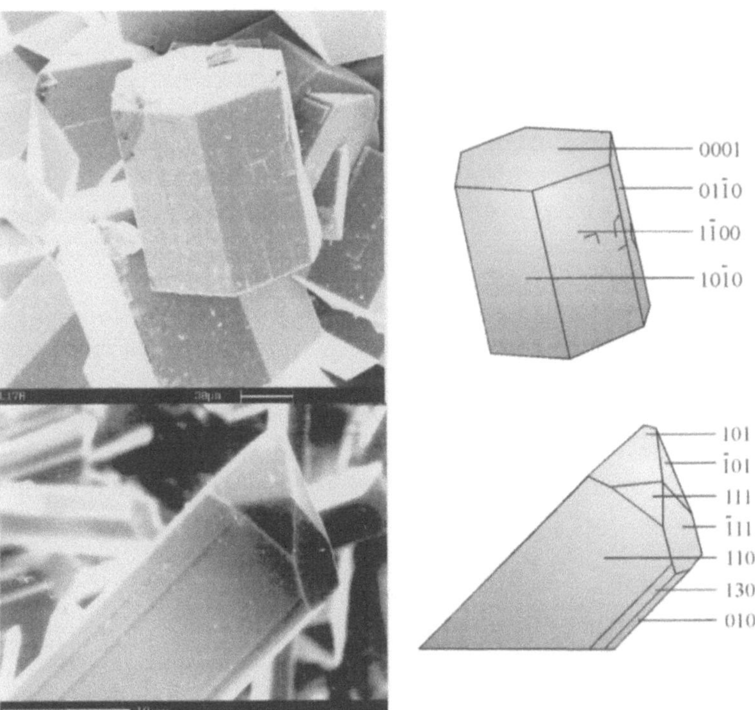

Figure 3.4: Effect of ageing on the prevailing phases present (same product stoichiometry: $1Li_2O \times 1Al_2O_3 \times 2SiO_2$).
Below: zeolite ABW (orthorhombic) obtained from glass precursors of the same composition at 280°C synthesis temperature, 42 days, 1kbar water pressure.
Above β-eucryptite (hexagonal) obtained from glass precursors of the same stoichiometry as zeolite ABW at 350°C synthesis temperature, 42 days, 1kbar water pressure.

Once the precursor is consumed by dissolution and the level of over-saturation approaches the equilibrium saturation C_s, particle growth can proceed by growth of large particles and simultaneous disappearing of the small ones by dissolution. This well-known effect in aqueous solutions is due to a minimization of the surface energy. The Gibbs-Kelvin equation (6) shows that for a certain level of over-saturation only one particle radius is at equilibrium. The mass transport process via the solution by dissolution-crystallization of crystalline phases can be very slow, but observable under the hydrothermal conditions (example: hydrothermal synthesis of quartz single crystals using polycrystalline quartz precursors).

The surface energy reduction can also be obtained by particle agglomeration; this process is particularly important under conditions near the point of zero charge, because charged surfaces tend to repel each other. However, under the special hydrothermal conditions at high pressure the classical particle behaviour in aqueous solutions is modified decisively.

3.1.4 Concluding Remarks

This mostly qualitative discussion emphasizes that many parameters have to be taken into account to understand hydrothermal reactions. A theoretical description of this complicated system is far from simple. Nevertheless, a few important conclusions can be drawn:

1. The main principle of the zeolite reconstruction work is to use the higher solubility of the metastable glassy precursor to trigger nucleation of the stable crystalline zeolite.
2. If the over-saturation is carefully maintained at a low level after the initial nucleation step, a narrow size distribution of the zeolite crystallites can be obtained.
3. The interface energy of the product is another important parameter that can lead to controlled heterogeneous nucleation, particularly on substrates where the zeolite/substrate interface energy is low.

3.2 Experimental

In order to perform long term experiments under elevated water pressure a special experimental set-up has to be employed permitting to maintain stable temperatures in tight sealed autoclaves, as temperature fluctuations lead to pressure fluctuations and, therefore, to non-reproducible changes of the physical and chemical processes during the synthesis. The autoclaves have to be kept sealed even at elevated pressures and temperatures and this for several weeks. A high pressure pump has to be installed able to adjust the chosen pressure at any time. The experimental arrangement is shown in figure 3.5.

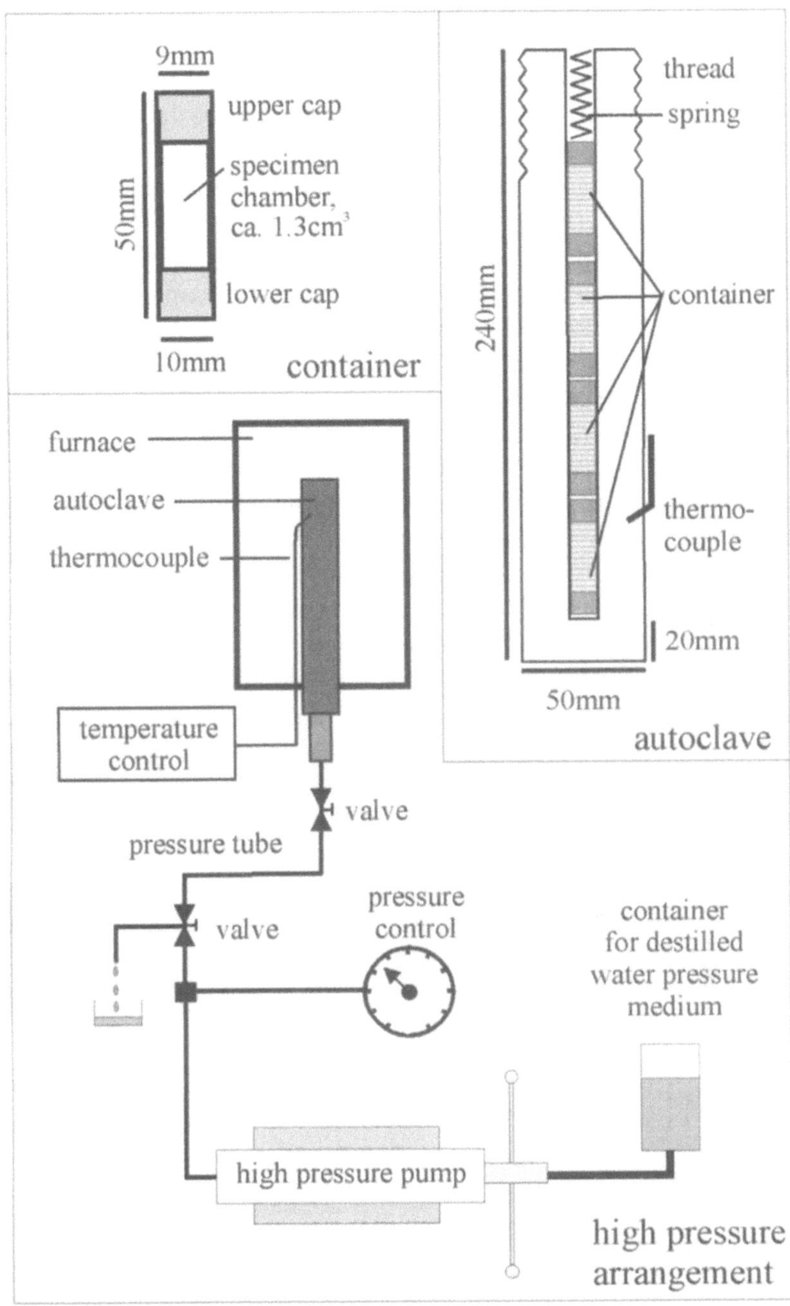

Figure. 3.5: Experimental set-up for the hydrothermal synthesis of zeolites under high pressure (between 0.5 and 2kbar, typically 1kbar). Left, above: Container made of Teflon, pure copper or pure gold. Right, above: Container arrangement within the autoclave unit (RMA-112R, TEM PRESS). Below: The experimental arrangement in order to apply an external water pressure.

After the temperature for the respective experiment is stabilized within the autoclave, the pressure is adjusted to the desired value (static conditions). Typical synthesis pressures for the hydrothermal experiment lay between 500 bar and 2 kbar, synthesis times in the zeolite reconstruction work between 14 and 60 days.

All autoclaves do have a characteristic temperature gradient. Four to eight autoclaves run in parallel, in order to cover the whole possible temperature range of a respective zeolite synthesis, as well as to reproduce the experiment simultaneously, without loosing valuable time.

Synthesis results essentially depend on the correct choice of synthesis pressure, temperature and time. Although a direct precursor glass – zeolite product relation exists, the educt stoichiometry in some cases has to be adapted in order to obtain the desired zeolite (see chapter 5).

Figure 3.6 shows the development of crystal morphology but also crystal symmetry of wairakite (ANA) with synthesis temperature: the higher the temperature, the higher the symmetry, an effect which can be correlated with investigations on natural phases, where the higher the symmetry, the higher the Si/Al disorder is found, as cubic symmetry is obtained at maximum randomness of Al-distribution in the framework[1]. Higher synthesis pressure only shifts the resulting stability ranges of wairakite to lower temperature values (see references on syntheses, chapter 5.2.1.2).

However, this phenomenon is mainly observed on zeolites with relatively high framework densities, i.e. very small channel systems. Zeolites with medium to big sized channel systems tend to change the crystal morphology with temperature (guarding the face normal angles) rather than to change crystal symmetry (see figure 3.7). Therefore, in the hydrothermal experiment under pressure crystalline zeolite phases are found having different crystal morphologies, dependent on the temperature of the synthesis. Investigations of these types of zeolites by X-ray diffraction phases reveal the same crystallographic symmetry[26].

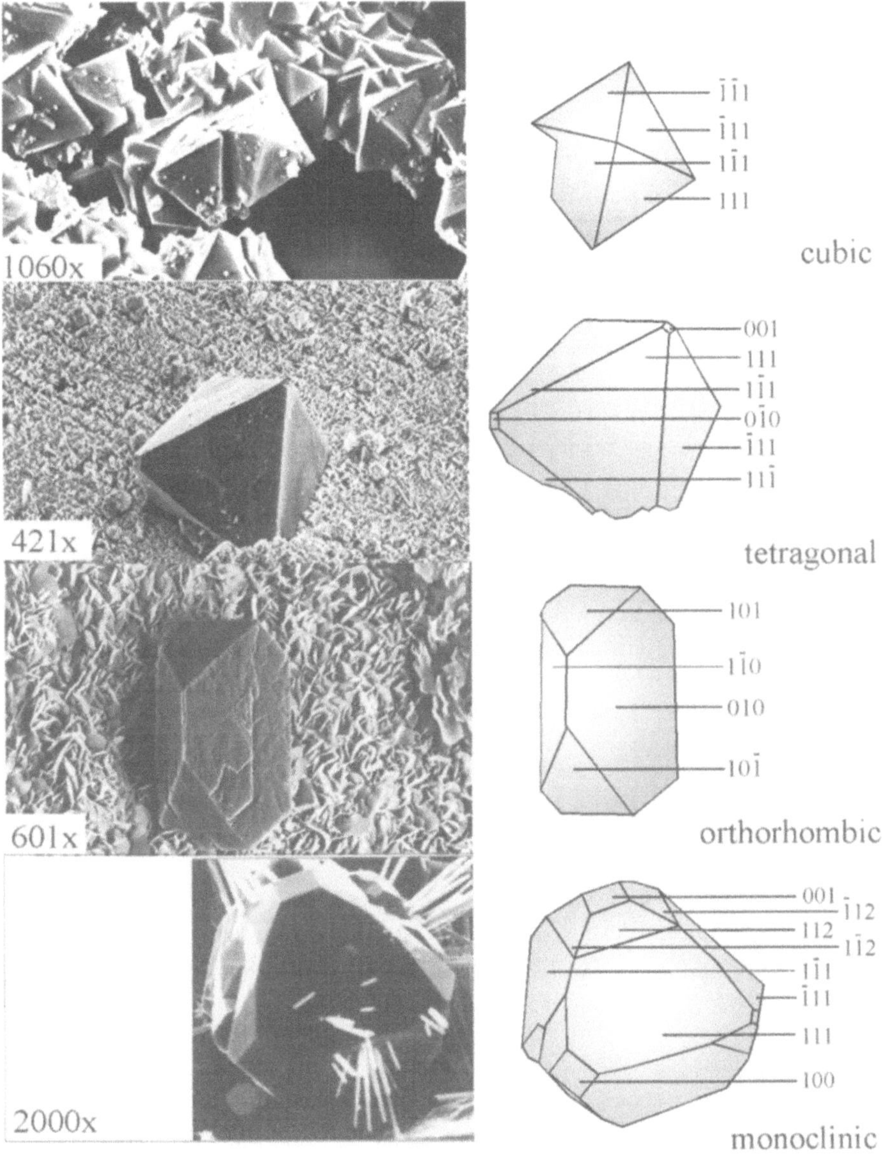

Figure. 3.6: The synthesis of zeolite Wairakite (ANA) by the hydrothermal method under pressure: Evolution of crystal morphology and symmetry with synthesis temperature starting with glass precursors of wairakite composition (42 days, 1kbar, different temperature levels): Monoclinic at 200°C, orthorhombic at 250°C, tetragonal at 320°C, cubic at 500°C.

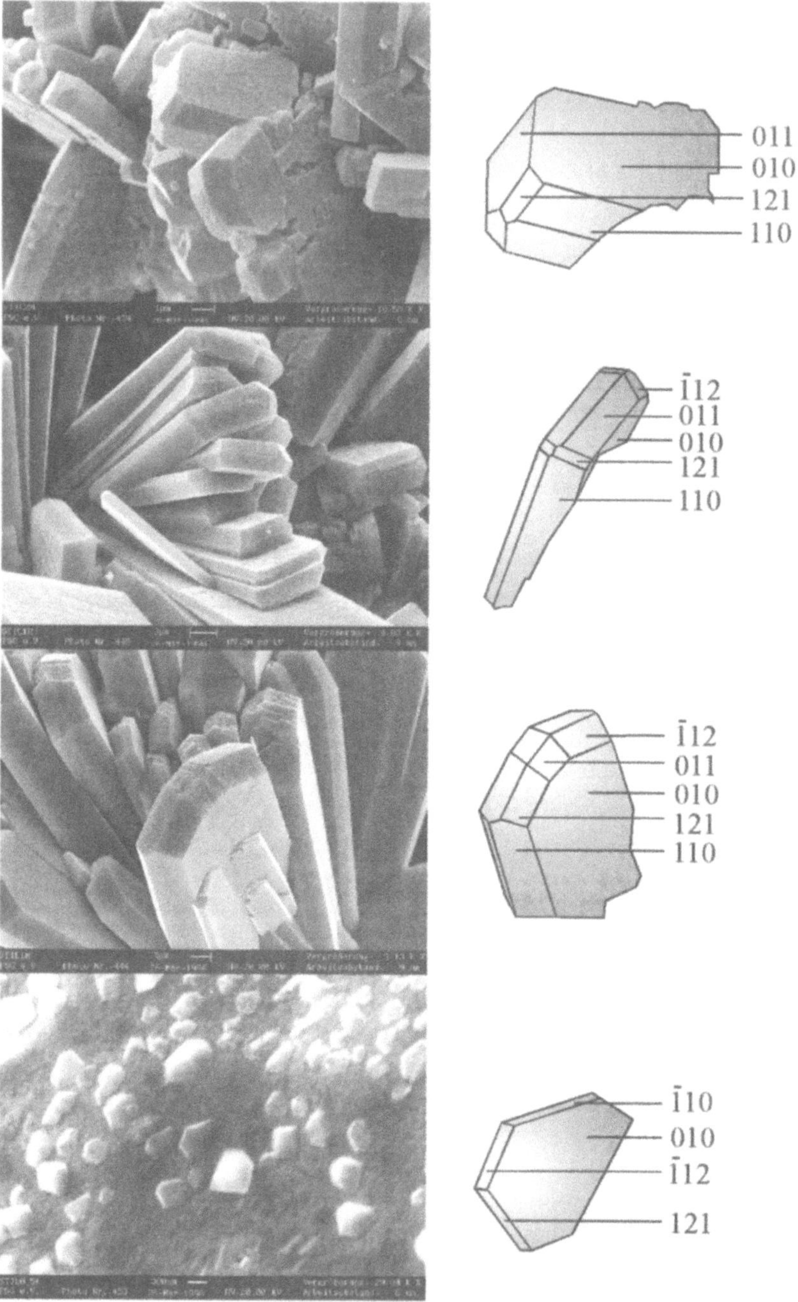

Figure 3.7: The development of zeolite Stilbite (STI) morphology (monoclinic, C2/m) during the hydrothermal synthesis under pressure at different temperatures (42days, 1kbar, from below to above): 155°C, 230°C, 250°C, 350°C.

Natural zeolites are mostly found in regions of former or present magmatic activity which means that high temperature and the presence of water influenced the formation. During the cooling process of a magmatic intrusion, the dissolved reminding ions, water-vapor, carbon dioxide and other volatile compounds are separated trying to reach the earth surface. This volatile phase also called fluid phase crystallizes during the cooling process loosing its high pressure. Different stages of crystallization can be observed during the decrease of temperature and pressure[27]. Only at the lowest stage of pressure and temperature, at the mildest conditions called the hydrothermal state, the formation conditions for alumino-silicate zeolite phases are reached. This can be achieved in nature not only by crystallization of the volatile phase itself in cavities, but also by intrusion into neighboring alumino-silicate beds transforming the material partially to zeolites. Even at the lowest pressures, like e.g. in hot springs (a phenomenon which is also related to magmatism) the formation of zeolites can be observed[1].

Natural zeolites are not only formed in the last stage of postmagmatic activity but also in the earliest stage of metamorphosis, the rock forming process, from alumino-silicate bearing sediments. Older metamorphic classifications used the term «laumontite facies» and also «zeolite facies» for the first step of metamorphosis, as zeolites, especially laumontite (Ca-zeolite), were the first new phases formed. At higher pressures and temperatures, zeolites are destroyed and denser alumino-silicate phases without zeolite properties are formed[28].

Low temperature and low pressure formation conditions in the sense of rock formation under the influence of water are the reason for the low density framework structure with characteristic channels and voids. Water molecules are an active element during this formation process, as they are the solvent for the ionic species forming the later zeolite and part of the zeolite structure under very special, « zeolitic » binding conditions.

Our approach was the simulation of these natural formation conditions of zeolites, therefore our new alternative synthesis method was first tested by the reconstruction of natural zeolites. Nature itself does neither use sol-gel processing, template molecules nor solutions containing high pH values for zeolite formation. In nature the formation process of zeolites is a process of dissolution of aluminosilicate or other, analogous, framework ion bearing oxide materials with water as dissolution and reaction medium under the influence of - essentially - temperature, pressure and time.

Consequently the zeolite phases synthesized with a specific ideomorphous morphology of crystals in a certain temperature interval, are directly indicating the temperature of their natural formation conditions at the given synthesis pressure.

4.   ZEOLITE   IDENTIFICATION   BY   THE   STEREO-COMPARATOR METHOD

General Considerations

X-ray diffraction is the first and standard method commonly used for the identification of zeolite phases. The identification of micro-crystalline phases by this method, however, is affecting two principal, general, problems. First the overlapping of diffraction peaks, especially in the case of zeolites, is complicating the further evaluation such as phase determination. Secondly, the synthesis product might be limited in quantity so that it is insufficient to be identified by the method.

One of the superior advantages of the hydrothermal synthesis method under high pressure is that conducting the experiments in an optimum way, micro-crystalline ideomorphouse crystal phases are obtained showing the characteristic morphology of typical crystals of the phase under investigation.

Ghobarkar [29,30], therefore, developed a new method for the identification of micro-crystals which allows the optical identification of crystals visualized by the scanning electron microscope. In contrast to the optical reflection goniometer, this method allows the measurement of crystal faces even in the micrometer range applying the crystallographic principle that the face normal angles of crystals keep constant independent from size. The face normal angles of an idiomorphouse crystal phase, however, are characteristic for each crystallographic system while the axis ratios are determined. Furthermore, the calculated axis ratios can be compared to X-ray diffraction data.

Tridimensional perception is based on the effect that the distance between the two eyes creates two different visual perspectives on the retinas. The optical nerve consisting of nerve fibers connects the retina of each eye to different halves of the brain, the respective optic chiasm. It controls the view direction, lens, pupil contraction etc. From there, the nerve fibers lead to the back of the head, to the occipital lobe. Only here, the recorded image is synthesized to a tridimensional picture of the observed objects taking into account the contributions obtained from the other half of the brain by specific interconnections (see figure 4.1).

Tridimensional objects appear in different contexts with respect to the two eyes and a comparing forward section is arising between the point F, fixed by the eyes, and other points before or behind. Tridimensional vision is, therefore, not absolute but relative. The differences in depth created by object points appearing in different spherical distances with respect to both eyes are called parallaxes. Ghobarkar could show that these parallaxes can be used to quantify the relative position of a plane of a microcrystal's face relative to the next. This is done in order to obtain all angles between the appearing faces (represented by their face normal angles).

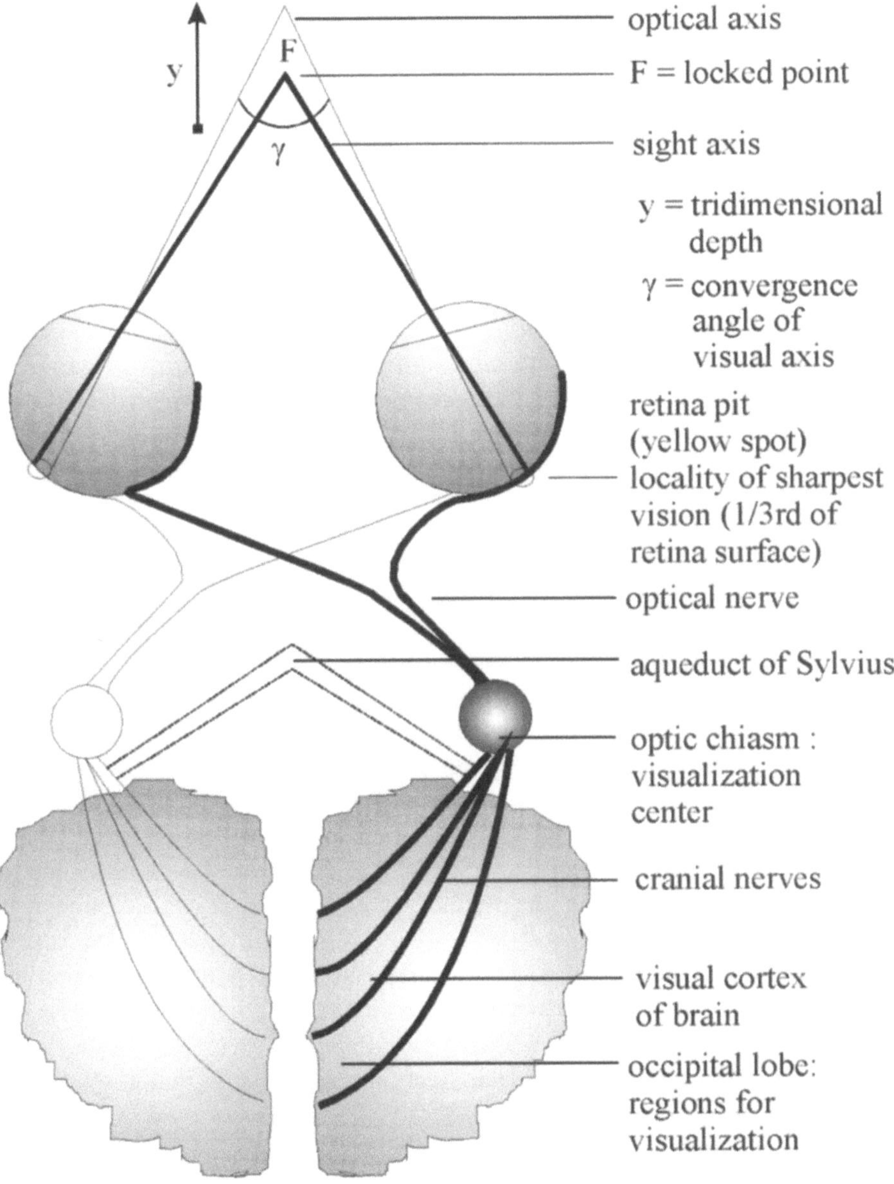

Figure 4.1: Physiological processes during stereoscopic viewing.

By using stereo scanning electron micrographs, crystals can be indexed and their crystallographic grouping is determined with the help of a stereo-comparator. Simultaneously, the energy dispersive X-ray (EDX) method allows the measurement of the chemical composition in a semi-quantitative way: Although the chemical composition of a crystalline phase may correspond to different minerals, the combination with the crystallographic properties obtained by the stereo-comparator method allows the determination of the crystal under investigation without any doubt.
The two different results are based on standard measurements in chemical composition and face angles.

4.1 Imaging Technique with the Scanning Electron Microscope

The stereo comparator method can be subdivided into different parts. In the electron-microscopic part, the crystalline phase under investigation is analysed by stereo-imaging. The specimen containing the micro-crystals is installed on the goniometer specimen stage of the scanning electron microscope. In a first approximation, the scanning electron microscope delivers parallel projection images of the observed objects. Although the parallaxes are object points appearing in different spherical distances with respect to both eyes, treating them as parallel projections is a good approximation.

Different perspectives for the stereo-comparator processing are created by taking two different images, the first at a position of $0°$ and the second after an inclination of $12°$. To get useful results, the inclination has to be done precisely in the same crystallographic zone. Figure 4.2 is illustrating the principle. Two different image pairs are taken in order to reduce systematic errors introduced by mechanical movement of the specimen stage (figure 4.3). It is important that the images are taken at the same value of magnification. Generally, the method is useful for crystals which need magnifications higher than 500 times as crystals bigger in size can be analysed by other methods. The smaller the crystals are, the higher the precision of the final phase angle measurements.

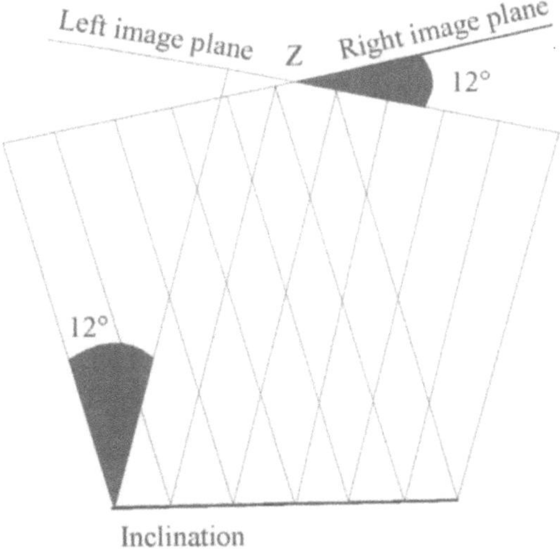

Figure 4.2: The principle of stereo-imaging with object thought resting: image planes and direction of projection confine the angle of inclination. The optimum inclination angle of 12° has been found empirically.

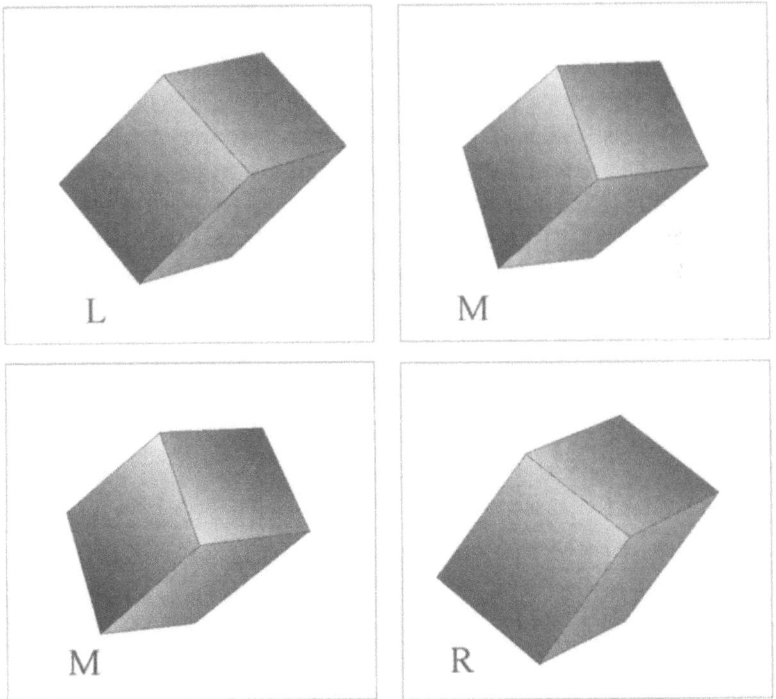

Figure 4.3: The position of the crystal images after inclination: L: −12° inclination, M: 0°, R: +12° (two pairs for control and accuracy purposes).

4.2 Image Evaluation with the Stereo-Comparator

The image paires obtained by the special SEM imaging technique are then placed into the stereo-comparator where they are merged into a tridimensional picture (see figure 4.4).

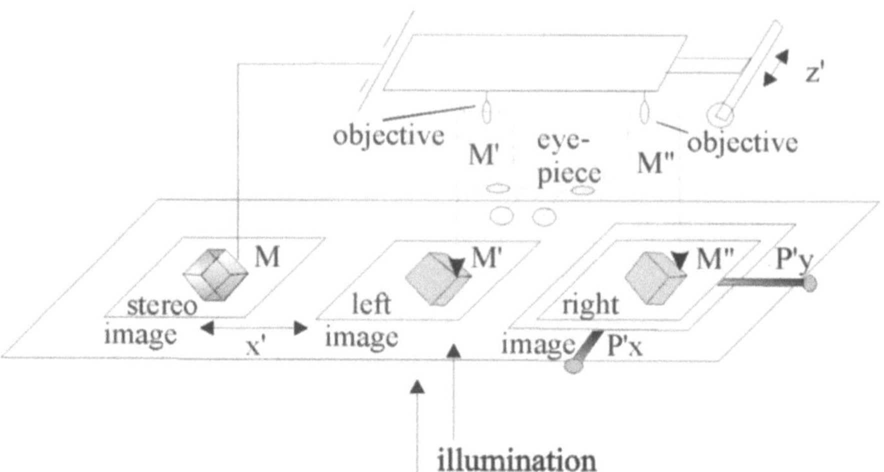

Figure 4.4: Schematic representation of the stereo-comparator.

The stereo-comparator gives the possibility to measure orthogonal coordinates x, y and their respective coordinate parallaxes Px and Py of an object and, therefore, to determine all its dimensions. Although the distance between the eyes is fixed and therefore the visual perspective, the telescope mechanism of the stereo-comparator allows to multiply the base distance and consequently the accuracy of the parallax-measurement (figure 4.5). A measuring counter assembled by the two parts of the optical device allows to determine the parallax-values in x and y by respective shifting.

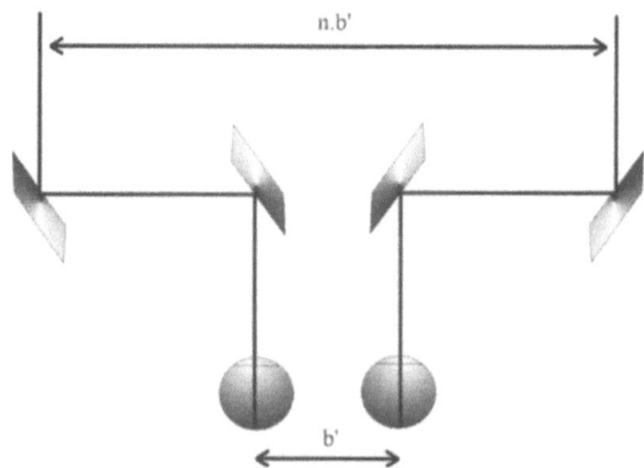

Figure 4.5: Means to increase the accuracy of distinction of depth by multiplication of the base distance b' by magnification.

4.3 The Calculation of x,y,z from Measured x,y, Px and Py.

The calculation of the face angles is done by determination of x,y as well as the parallaxes Px and Py for a respective point on a crystal face. Four points (three points to define a plane, plus one control point) are measured per crystal face. The co-ordinates x and y can be directly taken, while Py has to be kept constant carefully during the measurement in order to guarantee the accuracy. The z value for the respective point is calculated by:

$$z = Px/\sin 2\sin \tfrac{1}{2} \qquad (7)$$

Given that Px for both directions of inclination (-12°, 0°, 12°) give the same value (control of accuracy).

Doing this for three points (one supplementary point for control) a plane is clearly defined; the common form of the equation of a plane is:

$$Ax + By + Cz + D = 0 \qquad (8)$$

The angle between two planes 1 and 2 (crystal faces) is then given by:

$$\cos\alpha = A_1A_2 + B_1B_2 + C_1C_2/ \sqrt{(A_1^2 + B_1^2 + C_1^2)\,(A_2^2 + B_2^2 + C_2^2)} \qquad (9)$$

The calculus is rather simplified by using the vector form of the plane equation. This has the big advantage that the angle between two crystal faces is identical to the angle between their normal vectors. The determination of the angle between two faces covers therefore two steps.

First, the determination of the normal vectors of both planes:

The determined three points of a plane permit to calculate two vectors which pass within the plane. The normal vector of these planes is placed perpendicular to the plane and is the complementary angle to 180°.

Second, the determination of the angles between the normal vectors:

These are the angles between the crystal faces (see figure 4.6) obtained by the cross product of the two vectors.

Interested readers may find further details in a special textbook on the stereo-comparator method, to be published in 2004.

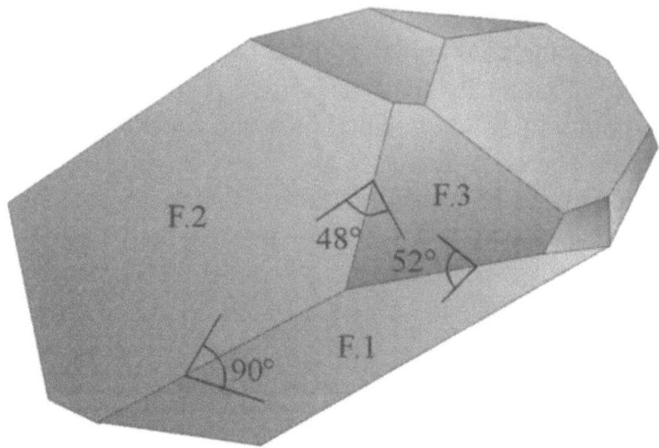

Figure 4.6: The angles between the crystals faces obtained by determing the face normal angles from the respective plane vectors for each face.

4.4 Crystal Indexing

In order to confirm the results on the face normal angles obtained by the stereo-comparator with respect to the crystal habit (crystal morphology), the values are written in the stereographic projection. At last, the stereographic projection has to be turned in such a way, that a standard set-up is achieved. The final indexing has to be accomplished by trial and error, while theoretical values can be taken into account once the crystal axis ratios and the crystal axis angles have been determined. Figure 4.7 shows the stereographic projection and the obtained indexed crystal at the end of the process.

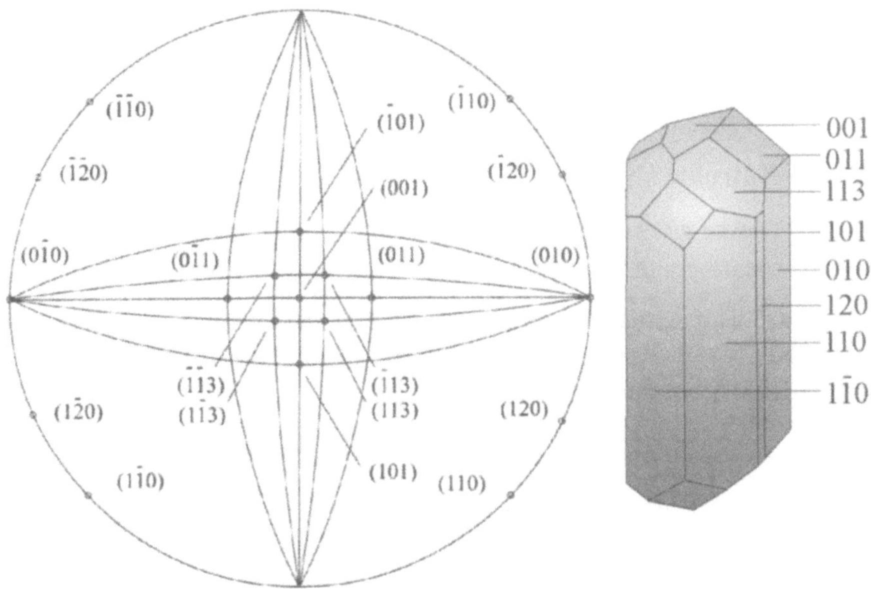

Figure 4.7: Left: Stereographic projection of Gonnardite (NAT). Right: The indexing of Gonnardite based on the stereographic projection (see also chapter 5.1.1.4).

5. THE ZEOLITE MINERALS

5.1 Zeolites with 4=1 Chains

5.1.1 Natrolite Group

5.1.1.1 Natrolite I $Na^+_{16}(H_2O)_{16}$ I $[Al_{16}Si_{24}O_{80}]$ – **NAT**: type material

> Name given by the sodium content (from Greek νατροσ - sodium bearing and λιδοσ = stone).

Remarks: The status of Paranatrolite and Tetranatrolite is not completely clarified[1,31,32,33]. However, we consider both as different grades of hydration of common natrolite.

Luster: silky

Channel system(s): <100> **8** 2.6 x 3.9** ↔ [001] **8** 2.5 x 4.9*

Framework density: 17.8 T/ nm^3

Cages/cavities: only given by the channel intersections

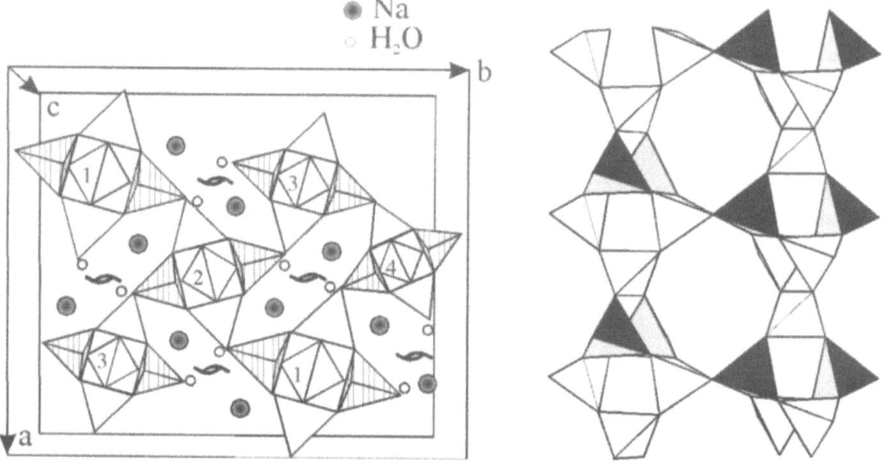

Figure 5.1: Projection of the Natrolite crystal structure [34,35,36,37].
Left: Arrangement of the fibrous zeolite chains viewed from c-axis. The tetrahedra positions are given in fractions of (-c/4), $[AlO_4]$-tetrahedra are shadowed. Na ions and H_2O molecules do have a characteristic arrangement in the channels of the alumino-silicate framework guarding the symmetry given by the 2_1 screw axis. The main channel system is directed parallel to c-axis.
Right: Connection of the alumino-silicate chains in the Natrolite structure. Note the arrangement of the alumina tetrahedra. The second intersected channel system perpendicular to c is formed by the windows left by the chain interconnection.

Cleavage: along (110)
Color: white, yellow, reddish
Crystallographic data: orthorhombic Fdd2
$$a = 1.83 \text{ nm}, b = 1.863 \text{ nm}, c = 0.66 \text{ nm}$$
Hardness: 5 - 5.5
SBU(s): 4=1

Synthesis conditions [38]:

High pressure hydrothermal treatment of synthetic glasses of Natrolite composition ($8Na_2O$ x $8 Al_2O_3$ x $24 SiO_2$).
Hydrothermal parameters: 60 days synthesis time, 1000 bar water pressure, temperature interval of isothermal conditions: 80°C - 200°C.

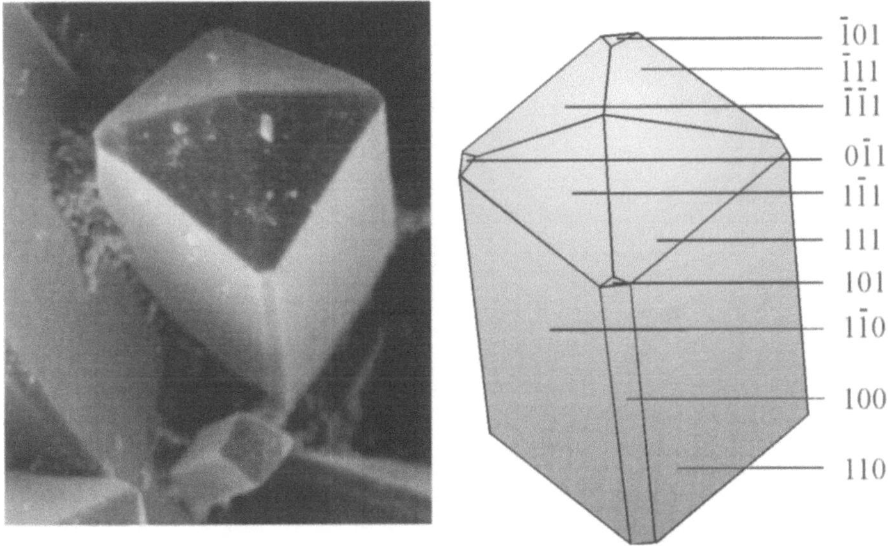

Figure 5.2: Morphology of synthetic Natrolite.
Left: SEM micrograph (200 X) of the as synthesized Natrolite.
Right: Indexing of the Natrolite crystal confirming the orthorhombic symmetry. Note that the zeolite channels are linked to the main crystallographic directions [001] and [110].

Other synthesis approaches: See [39,40,41].

5.1 Zeolites with 4=1 Chains

5.1.1 Natrolite Group

5.1.1.2 Mesolite I Na^+_{16} Ca^{2+}_{16} $(H_2O)_{64}$ I $[Al_{64}Si_{72}O_{240}]$ - NAT

Name from Greek μεσον = middle and λιδοσ = stone.

Luster: silky

Channel system(s): <100> **8** 2.6 x 3.9** ↔ [001] **8** 2.5 x 4.9*

                            dependent on the actual chemical composition

Framework density: 17.8 T/ nm^3

Cages/cavities: only given by the channel intersections

Cleavage: along (110)

Color: white

Crystallographic data: orthorhombic Fdd2

                          a = 1.84 nm, b = 5.665 nm, c = 0.665 nm

Hardness: 5 - 5.5

SBU(s): 4=1

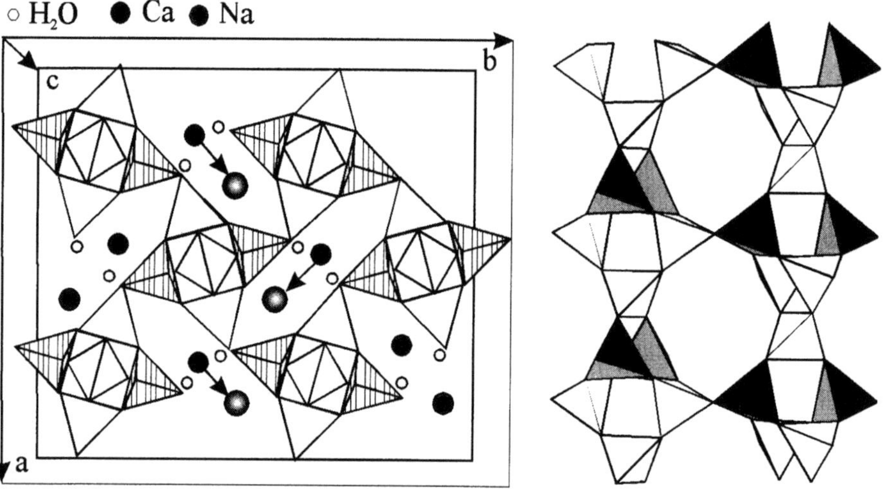

Figure 5.3: Projection of the Mesolite crystal structure [42,43,44].

Left: Arrangement of the fibrous zeolite chains viewed from c-axis. The tetrahedra positions are identical as those in the Natrolite structure, $[AlO_4]$-tetrahedra are shadowed. Two Na ions are replaced by one Ca - ion taking a former Na - position as indicated by an arrow. The main channel system is directed parallel to c-axis.

Right: Connection of the alumino-silicate chains in the Mesolite structure, in an identical way as for Natrolite. The second intersected channel system perpendicular to c is formed by the windows left by the chain interconnection.

Synthesis conditions [45]:

High pressure hydrothermal treatment of synthetic glasses of Mesolite composition ($8Na_2O$ x $16CaO$ x $24Al_2O_3$ x $72SiO_2$).
Hydrothermal parameters: 60 days synthesis time, 1000 bar water pressure, temperature interval of isothermal conditions: 80°C - 200°C.

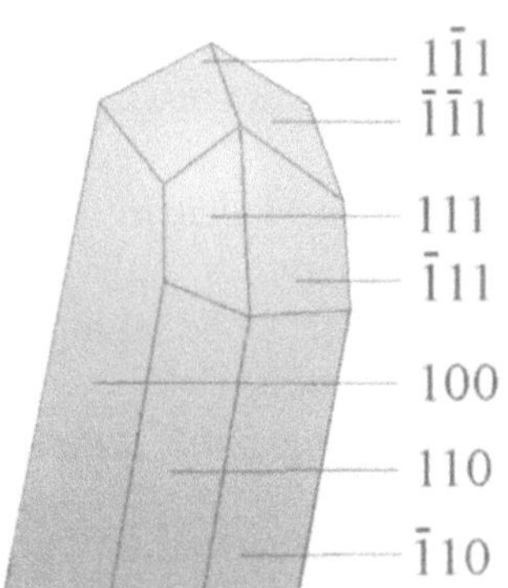

Figure 5.4: Morphology of synthetic Mesolite.
Left: SEM micrograph (5000 X) of the as synthesized Mesolite.
Right: Indexing of the Mesolite crystal confirming the orthorhombic symmetry. Note that the zeolite channels are linked to the crystallographic directions [001] and [110].

The authors are not aware of other synthesis approaches.

5.1 Zeolites with 4=1 Chains

5.1.1 Natrolite Group

5.1.1.3 Scolecite I $Ca^{2+}_8 (H_2O)_{24}$ I $[Al_{16}Si_{24}O_{80}]$ - NAT

Name from Greek σκωληζ = worm and λιδοσ = stone due to the behavior upon heating.

Luster: silky
Channel system(s): <100> **8** 2.6 x 3.9** ↔ [001] **8** 2.5 x 4.9*
                          dependent on the actual chemical composition
Framework density: 17.8 T/ nm^3
Cages/cavities: only given by the channel intersections
Cleavage: good, along (110)
Color: white
Crystallographic data: monoclinic F1d1
                    a = 1.851 nm, b = 1.897 nm, c = 0.653 nm
                    β = 90.65°
Hardness: 5 - 5.5
SBU(s): 4=1

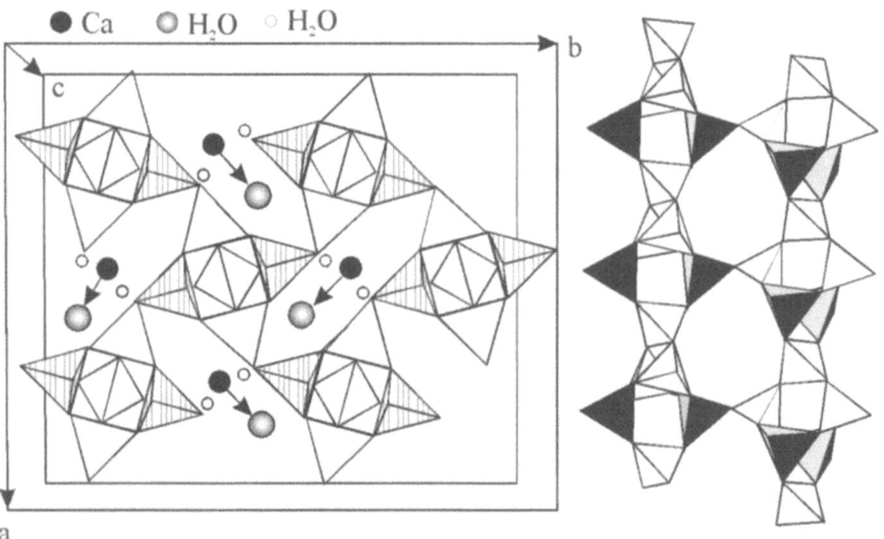

Figure 5.5: Projection of the Scolecite crystal structure [46,47,48,49,50].
Left: Arrangement of the fibrous zeolite chains viewed from c-axis, [AlO$_4$]-tetrahedra are shadowed. Only Ca ions are present in the main channel system which is directed parallel to c - axis.
Right: Connection of the alumino-silicate chains in the Scolecite structure comparable to Natrolite and Mesolite but distorted due to the lower symmetry. The second intersected channel system perpendicular to c is formed by the windows left by the chain interconnections.

Synthesis conditions [51]:

High pressure hydrothermal treatment of synthetic glasses of Scolecite composition ($8CaO \times 16Al_2O_3 \times 24SiO_2$).
Hydrothermal parameters: 60 days synthesis time, 1000 bar water pressure, temperature interval of isothermal conditions: 80°C - 200°C.

Figure 5.6: Left the SEM micrograph (1000 X) of the as synthesized Scolecite. Right: Indexing of the Scolecite crystal confirming the monoclinic symmetry. Note that the zeolite channels are linked to the crystallographic directions [001] and [110].

Other synthesis approaches: See [52,53].

5.1   Zeolites with 4=1 Chains

5.1.1 Natrolite Group

5.1.1.4 Gonnardite I Na^+_5 Ca^{2+}_2 $(H_2O)_{12}$ I $[Al_9Si_{11}O_{40}]$ - NAT

Named after the Frenchman F. Gonnard, earliest descriptor.

Luster: silky
Channel system(s): <100> **8** 2.6 x 3.9** ↔ [001] **8** 2.5 x 4.9*
                    dependent on the actual chemical composition
Framework density: 17.8 T/ nm^3
Cages/cavities: only given by the channel intersections
Cleavage: none
Color: white, yellowish
Crystallographic data: tetragonal I42d (Si,Al disordering) or orthorhombic
Pbmn (Si,Al ordering)
                    a = 1.35 nm, b = 1.35 nm, c = 0.665 nm
Hardness: 5 - 5.5
SBU(s): 4=1

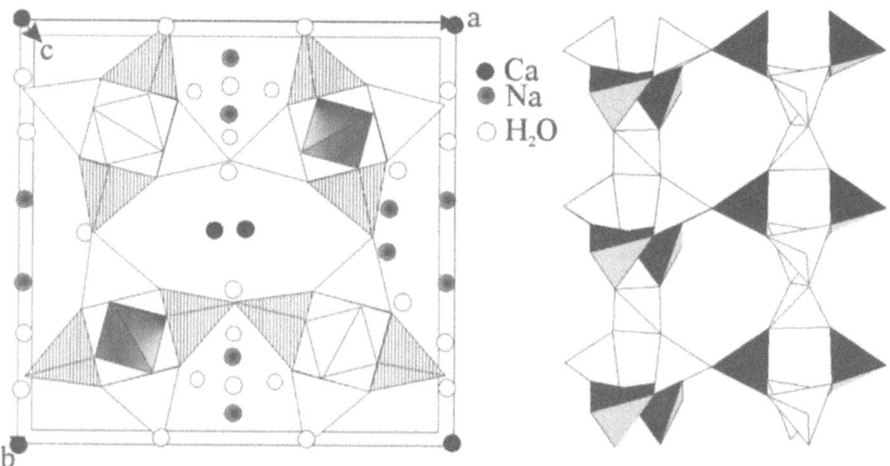

Figure 5.7: Projection of the Gonnardite crystal structure [54,55,56].
Left: Arrangement of the fibrous zeolite chains viewed from c-axis, [AlO4]-
tetrahedra are shadowed. Ca- and Na- ions are present in the main channel system
which is directed parallel to c - axis.
Right: Connection of the alumino-silicate chains in the Gonnardite structure
directly comparable to Natrolite and Mesolite. The second intersected channel
system perpendicular to c is formed by the windows left by the chain
interconnections.

Synthesis conditions [57]:

High pressure hydrothermal treatment of synthetic glasses of Gonnardite composition ($5Na_2O$ x $8CaO$ x $18Al_2O_3$ x $22SiO_2$).
Hydrothermal parameters: 14 days synthesis time, 1000 bar water pressure, temperature interval of isothermal conditions: 80°C - 150°C.

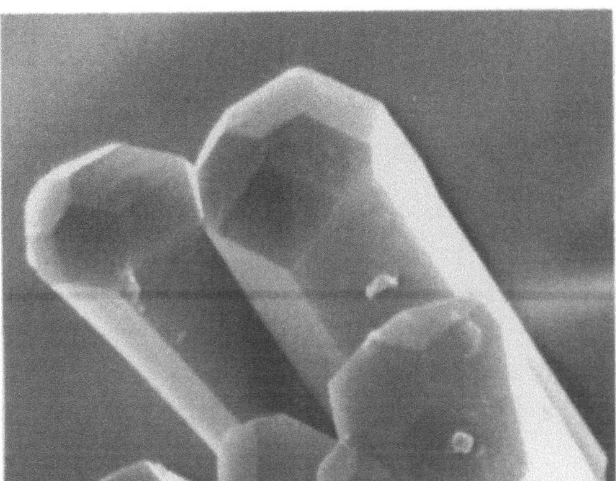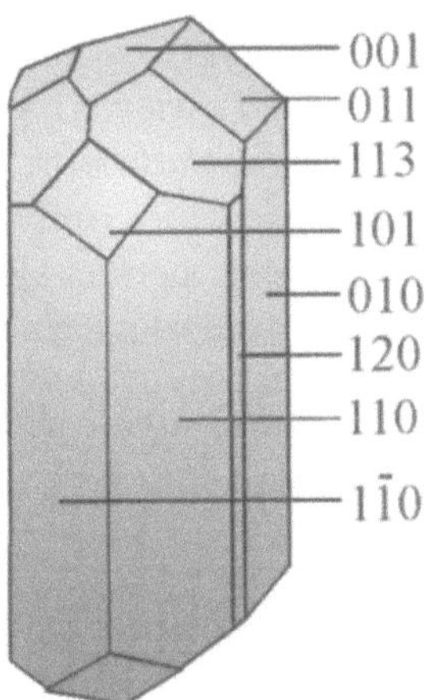

Figure 5.8: Left the SEM micrograph (2750 X) of the as synthesized Gonnardite. Right: Indexing of the Gonnardite crystal revealing the orthorhombic symmetry. Note that the zeolite channels are linked to the main crystallographic directions [001] and [110].

Other synthesis approaches: See [58].

Zeolites with 4=1 Chains

5.1.2 Edingtonite Group

5.1.2.1 Edingtonite | $Ba^{2+}_2(H_2O)_8$ | $[Al_4Si_6O_{20}]$ – EDI: type material

Named after Mr. Edington, mineral collector from Glasgow.

Luster: silky
Channel system(s): <110> **8** 2.8 x 3.8** ↔ [001] **8** 2.0 x 3.1*
Framework density: 16.6 T/ nm^3
Cages/cavities: only given by the channel intersections
Cleavage: perfect along (111)
Color: colorless, transparent

Crystallographic data: tetragonal P$\bar{4}2_1$m
         a = 0.958 nm, b = 0.958 nm, c = 0.652 nm
         or orthorhombic P2$_1$2$_1$2$_1$
         a = 0.955 nm, b = 0.967 nm, c = 0.652 nm

Hardness: 4
SBU(s): 4=1

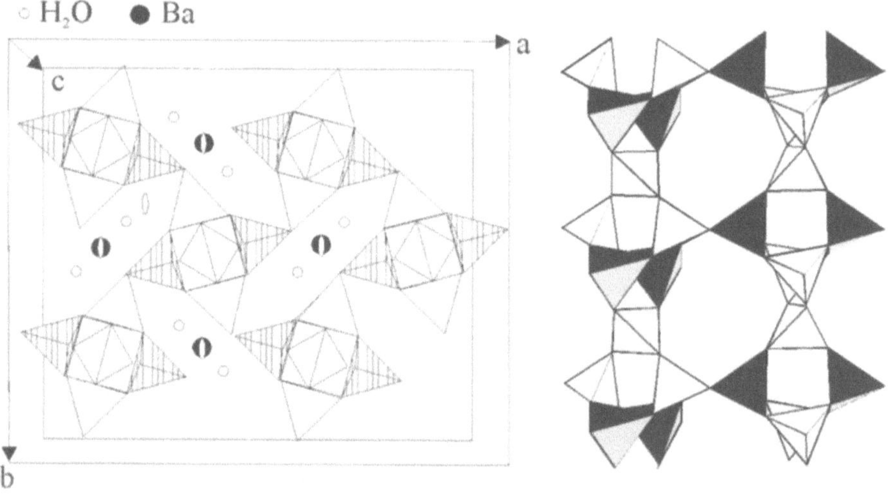

Figure 5.9: Projection of the Edingtonite crystal structure [59,60,61,62].
Left: Arrangement of the fibrous zeolite chains viewed from c-axis, [AlO$_4$]-tetrahedra are shadowed. Ba - ions are present in the main channel system which is directed parallel to c - axis.
Right: Connection of the alumino-silicate chains in the Edingtonite framework in analogy to the arrangement in the THO framework. Note the difference to the chain arrangement in NAT framework. The intersected channel systems perpendicular to c are formed by the windows left by the chain interconnections.

Synthesis conditions [63]:

High pressure hydrothermal treatment of synthetic glasses of Edingtonite composition ($2BaO \times 2Al_2O_3 \times 6SiO_2$).
Hydrothermal parameters: 28 days synthesis time, 1000 bar water pressure, temperature interval of isothermal conditions: 80°C - 230°C.

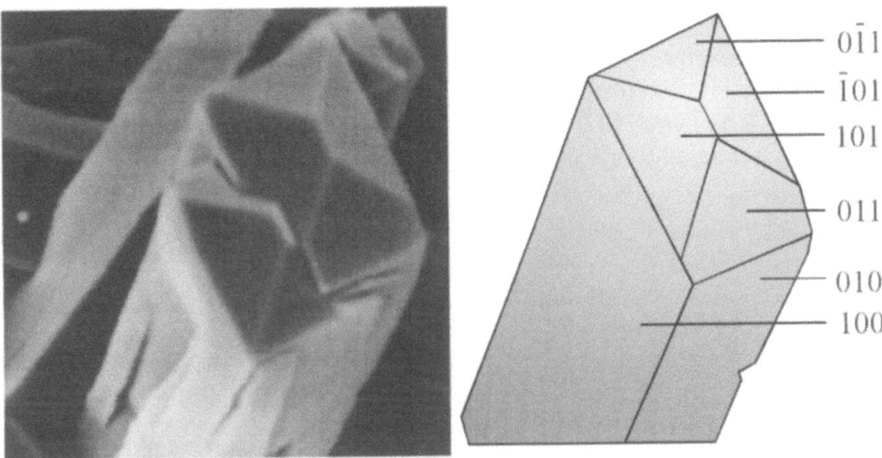

Figure 5.10: Left the SEM micrographs (5000 X) of the synthesized Edingtonite (twinned). Right: Indexing of the Edingtonite crystal revealing the orthorhombic symmetry. The zeolite channels are linked to the crystallographic directions [001] and [110].

Other synthesis approaches: See [64,65,66,67,68,69].

5.1 Zeolites with 4=1 Chains

5.1.3 Thomsonite Group

5.1.3.1 Thomsonite I $Na^+_4 Ca^{2+}_8 (H_2O)_{24}$ I $[Al_{20}Si_{20}O_{80}]$ – **THO**: type

material. Named after T. Thomson, English journal editor 19th C.

Luster: silky
Channel system(s): [100] **8** 2.3 x 3.9* ↔ [010] **8** 2.2 x 4.0* ↔
                [001] **8** 2.2 x 3.0*
Framework density: 17.7 T/ nm^3
Cages/cavities: only given by the channel intersections
Cleavage: perfect along (100), good along (010)
Color: colorless, yellowish
Crystallographic data: orthorhombic Pcnn
                a = 1.305 nm, b = 1.309 nm, c = 1.322 nm
Hardness: 5 - 5.5
SBU(s): 4=1

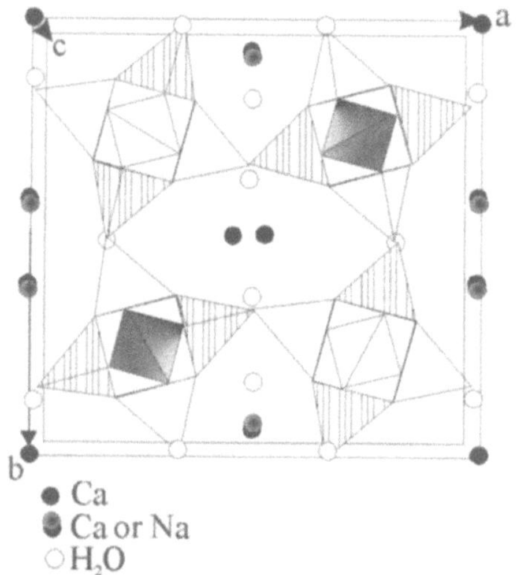

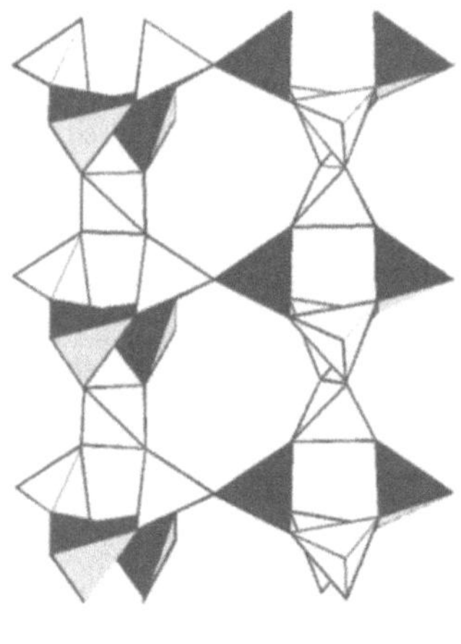

Figure 5.11: Projection of the Thomsonite crystal structure [70,71,72,73,74,75].
Left: Arrangement of the fibrous zeolite chains viewed from c-axis, $[AlO_4]$-tetrahedra are shadowed. Ca- and Na- ions are present in the main channel system which is directed parallel to c - axis.
Right: Connection of the alumino-silicate chains in the Thomsonite structure. Note the difference to the chain arrangement in NAT structure types. The intersected channel systems perpendicular to c are formed by the windows left by the chain interconnections.

Synthesis conditions [63]:

High pressure hydrothermal treatment of synthetic glasses of Thomsonite composition ($4Na_2O$ x $16CaO$ x $20Al_2O_3$ x $40SiO_2$).
Hydrothermal parameters: 14 days synthesis time, 1000 bar water pressure, temperature interval of isothermal conditions: 80°C - 230°C.

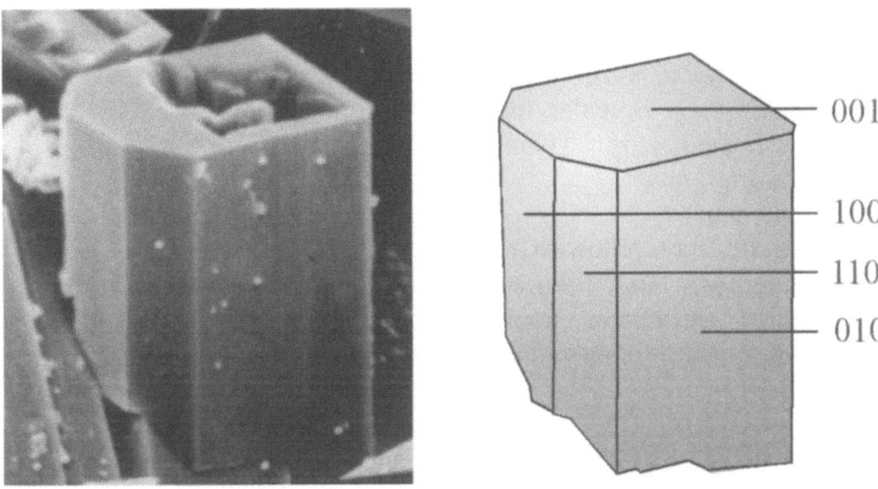

Figure 5.12: Left the SEM micrographs (2000 X) of the synthesized Thomsonite. Right: Indexing of Thomsonite crystal revealing the orthorhombic symmetry. The zeolite channels are linked to the main crystallographic directions [100], [010] and [001].

Other synthesis approaches: See [76,52,77,78,79,80,81].

5.2 Zeolites with Singly Connected 4-Ring Chains

5.2.1 Analcime group

5.2.1.1 Analcime I Na^+_{16} $(H_2O)_{16}$ I $[Al_{16}Si_{32}O_{96}]$ – **ANA**: type material

Name from Greek αναλχιμοσ = "without strength" - with respect to the weak electrostatic effects upon friction.

Luster: glassy, translucent
Channel system(s): <110> **8** 1.6 x 4.2* (see also chapter 2.3)
               Irregular, formed by distorted 8-rings
Framework density: 18.5 T/ nm^3
Cages/cavities: none
Cleavage: none
Color: white, gray, yellowish, reddish
Crystallographic data: from cubic (maximum framework cation disorder to triclinic (maximum framework cation order)[1] – dependent on formation/synthesis temperature[82]:

        triclinic (lowest temperature), monoclinic, orthorhombic, tetragonal, <u>cubic</u> (highest temperature): Ia3d

           a = 1.373nm

Hardness: 5.5
SBU(s): 4

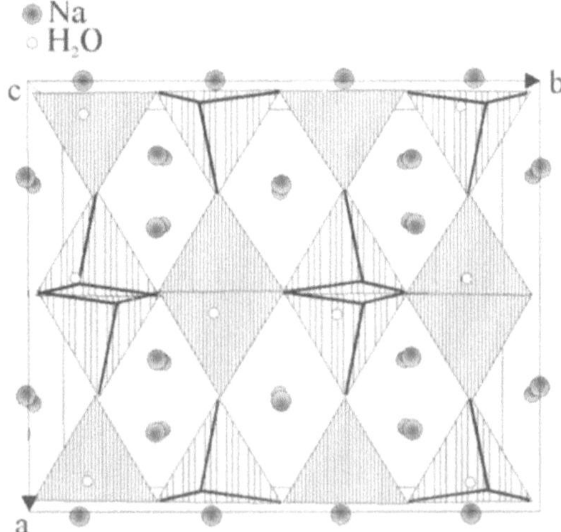

Figure 5.13: Projection of the Analcime crystal structure [83,84].
Arrangement of the four ring chains viewed from c-axis, $[AlO_4]$-tetrahedra are shadowed. The tetrahedra are only connected at the corners. Na- ions and water molecules are present in the vincinities as the channels are not visible.

Synthesis conditions [85,82]:

High pressure hydrothermal treatment of synthetic glasses of Analcime composition ($4Na_2O$ x $8Al_2O_3$ x $32SiO_2$).
Hydrothermal parameters: up to 60 days synthesis time, 1000 to 2000 bar water pressure, temperature interval of isothermal conditions: 80°C - 450°C.

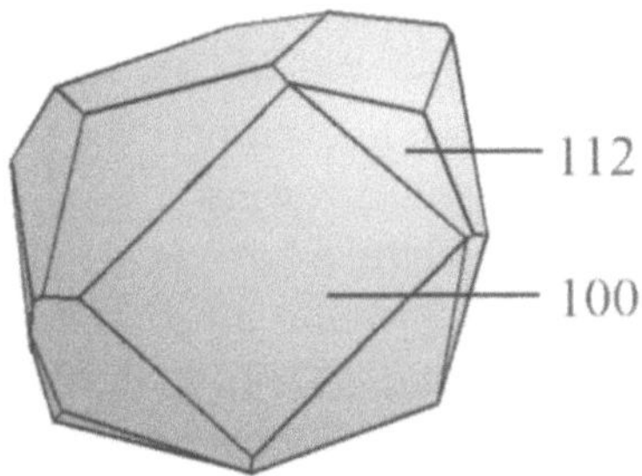

Figure 5.14: Left: the SEM micrographs (2000 X) of the synthesized Analcime (high temperature range). Right: Indexing of a respective Analcime crystal revealing the cubic symmetry. The very small zeolite channels pass along the crystallographic direction [110] without intersection.

Other synthesis approaches: See [86,87,88,89,90,91,92,93,94,52,95,96].

5.2 Zeolites with Singly Connected 4-Ring Chains

5.2.1 Analcime group

5.2.1.2 Wairakite | Na^+_{16} $(H_2O)_{16}$ | $[Al_{16}Si_{32}O_{96}]$ – ANA

Named after the locality Wairakei, New Zealand.

Luster: glassy, translucent
Channel system(s): <110> **8** 1.6 x 4.2* (see also chapter 2)
                     irregular, formed by distorted 8-rings
Framework density: 18.5 T/ nm^3
Cages/cavities: none
Cleavage: very weak
Color: white, colorless
Crystallographic data: from cubic (maximum framework cation disorder to triclinic (maximum framework cation order) – dependent on formation/synthesis temperature [97] (see Analcime):

> Monoclinic C 2/c
> a = 1.369nm, b = 1.364nm, c = 1.356nm
> $\beta = 90.5°$

Hardness: 5 - 5.5
SBU(s): 4

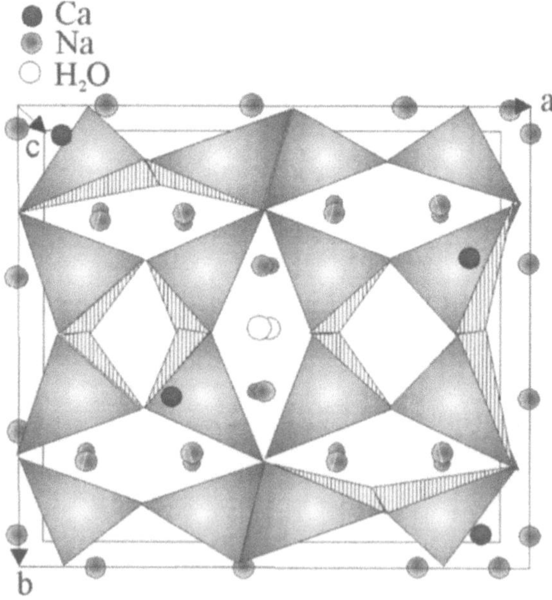

Figure 5.15: Projection of the Wairakite crystal structure [98,99].
Arrangement of the four ring chains viewed from c-axis, [AlO₄]-tetrahedra are shadowed. Note that the tetrahedra are only connected at the corners. Ca- ions and water molecules are present in the vincinities as the channels are not visible.

Synthesis conditions [100,97]:

High pressure hydrothermal treatment of synthetic, water free glasses of Wairakite composition (8CaO x 8Al$_2$O$_3$ x 32SiO$_2$).
Hydrothermal parameters: up to 60 days synthesis time, 1000bar – 2000bar water pressure, temperature interval of isothermal conditions: 85°C - 500°C.

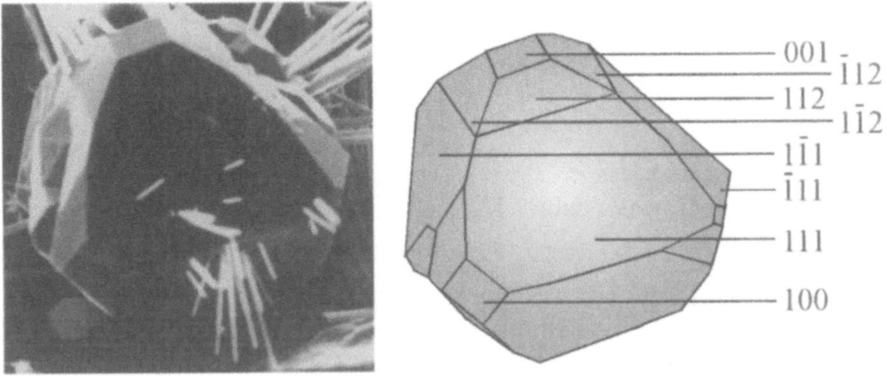

Figure 5.16: Left: The SEM micrographs (2000 X) of the synthesized Wairakite (low temperature range of synthesis). Right: Indexing of a monoclinic Wairakite crystal. The very small zeolite channels pass along the crystallographic direction [110] without intersection.

Other synthesis approaches: See [52,53,101,102].

5.2 Zeolites with Singly Connected 4-Ring Chains

5.2.1 Analcime group

5.2.1.3 Hsianghualite | $Li^+_{16} Ca^{2+}_{24} 8F_2$ | $[Be_{24}Si_{24}O_{96}]$ – ANA

Name from a Chinese word for fragrant flower

Luster: glassy, translucent
Channel system(s): <110> 8 1.6 x 4.2* (see also chapter 2)
                Irregular, formed by distorted 8-rings
Framework density: 18.5 T/ nm^3
Cages/cavities: none
Cleavage: none
Color: white, colorless
Crystallographic data: cubic, $I2_13$, a = 1.288nm
Hardness: 6.5
SBU(s): 4

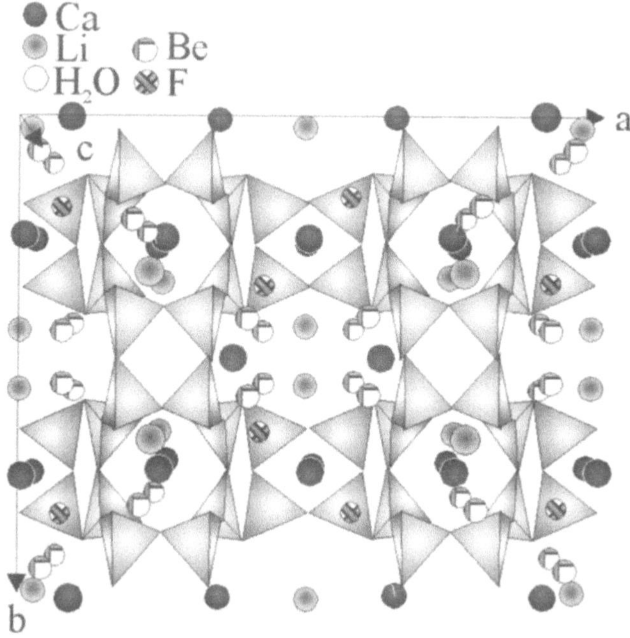

Figure 5.17: Projection of the Hsianghualite crystal structure [103].
Arrangement of the four ring chains with view into the distorted channels; [BeO$_4$]-tetrahedra are distributed with perfect Be, Si alteration as the content is equal. Tetrahedra are only connected at the corners.

Synthesis conditions [97]:

High pressure hydrothermal treatment of synthetic water-free glasses of Hsianghualite composition ($8Li_2O$ x $16CaO$ x $8CaF_2$ x $24BeO$ x $24SiO_2$). Hydrothermal parameters: up to 42 days synthesis time, 1000 bar water pressure, temperature interval of isothermal conditions: 150°C - 500°C.

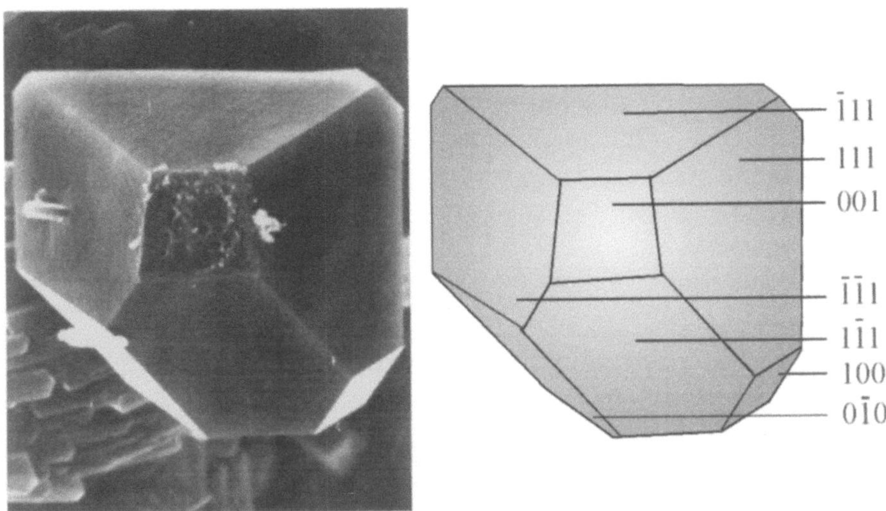

Figure 5.18: Left: The SEM micrograph (1000 X) of the synthesized Hsianghualite. Right: indexing of the Hsianghualite crystal of cubic symmetry.

Other synthesis approaches: See [104].

5.2 Zeolites with Singly Connected 4-Ring Chains

5.2.1. Analcime group

5.2.1.4 Viséite I Na^+_2 Ca^{2+}_{10} $(H_2O)_{16}$ I $[Al_{20}Si_6P_{10}O_{60}(OH)_{36}]$ – ANA

Named after the locality Visé, Belgium.

Remarks: officially discredited as zeolite[31] but without final proof[1,105,106,107].
Luster: glassy
Channel system(s): <110> **8** 1.6 x 4.2* (see also chapter 2)
                    Irregular, formed by distorted 8-rings
Framework density: 18.5 T/ nm^3
Cages/cavities: none
Cleavage: none
Color: white, bluish, yellowish

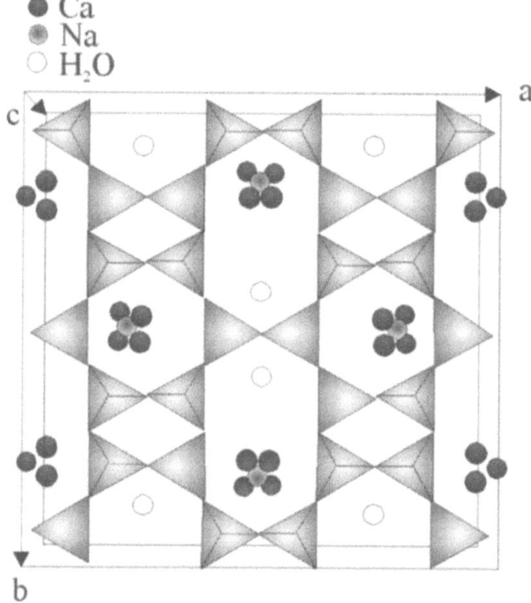

Figure 5.19: Projection of the Viséite crystal structure[1,107].
Arrangement of the four ring chains with view into the small channels; the ordering of $[PO_4]$-tetrahedra depends strongly on the synthesis/formation temperature.

Crystallographic data: from cubic (maximum framework cation disorder to triclinic (maximum framework cation order) – dependent on formation/synthesis temperature :
triclinic (lowest temperature), monoclinic- not observed, orthorhombic, tetragonal, <u>cubic</u> (highest temperature): I2$_1$3

$$a = 1.365nm$$

Hardness: 3-4
SBU(s): 4

Synthesis conditions [82]:

High pressure hydrothermal treatment of synthetic glasses of Viséite composition (1Na$_2$O x 10CaO x 10Al$_2$O$_3$ x 6SiO$_2$ x 5P$_2$O$_5$).
Hydrothermal parameters: up to 42 days synthesis time, 1000 bar – 2000bar water pressure, temperature interval of isothermal conditions: 80°C - 630°C.

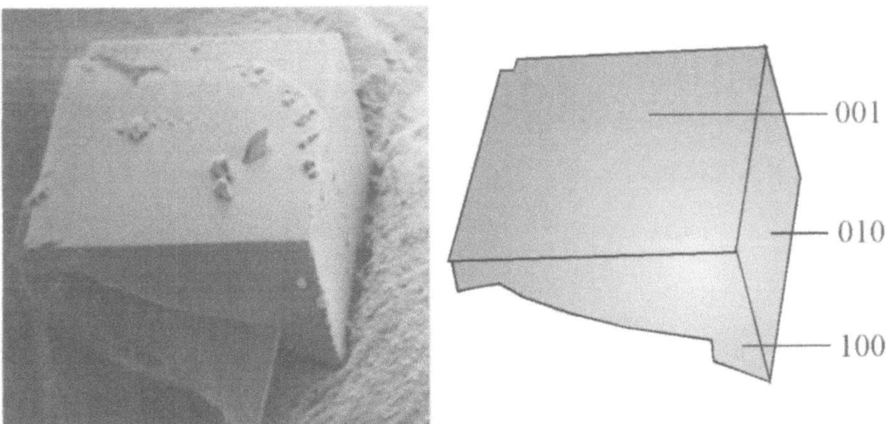

Figure 5.20: Left: The SEM micrograph (1550 X) of the synthesized Viséite (high temperature). Right: Indexing of the Viséite crystal revealing the cubic symmetry. The small distorted zeolite channels pass along the crystallographic directions [100] without intersections.

Other synthesis approaches: See [108].

5.2 Zeolites with Singly Connected 4-Ring Chains

5.2.2 Laumontite group

5.2.2.1 Laumontite I Ca^{2+}_4 $(H_2O)_{16}$ I $[Al_8Si_{16}O_{48}]$ – LAU: type material

Named after Gillet de Laumont, French mineralogist.

Remarks: Leonhardite is discredited as a proper zeolite since it is formed by simple water loss of Laumontite.

Luster: pearly on (110)

Channel system(s): [001] **10** 4.0 x 5.3* ↔ [100] **6*** ↔ [010] **6*** I [010] **6***
(size not determined: arrangement see chapter 2.3)

Framework density: 17.8 T/ nm^3

Cages/cavities: only given by channel intersections

Cleavage: according (010) and (100)

Color: colorless, yellowish, reddish

Crystallographic data: monoclinic, C2/m
a = 1.504nm, b = 1.317nm, 0.771nm
β = 113.2°

Hardness: 3-3.5

SBU(s): 4, 6 or 6-2

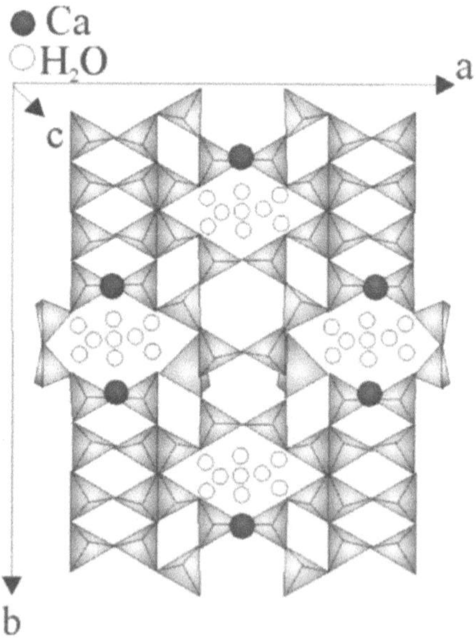

Figure 5.21: Projection of the Laumontite crystal structure [109,110,111,112].
Arrangement of the four ring chains viewed from c-axis, with view into the channel system. Ca- ions and water molecules are present.

Synthesis conditions [113]:

High pressure hydrothermal treatment of water free synthetic glasses of Laumontite composition ($4CaO$ x $4Al_2O_3$ x $16SiO_2$).
Hydrothermal parameters: 42 days synthesis time, 1000 bar water pressure, temperature interval of isothermal conditions: 30°C - 450°C. Absence of CO_2.

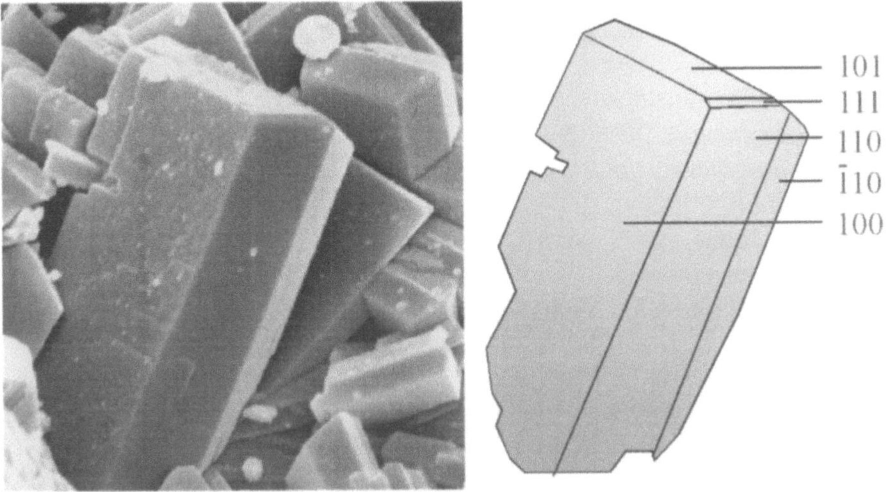

Figure 5.22: Left: The SEM micrograph (4640 X) of synthesized Laumontite Right: Indexing of Laumontite crystal revealing the monoclinic symmetry.

Other synthesis approaches: See [114,53,115].

5.2 Zeolites with Singly Connected 4-Ring Chains

5.2.3 Roggianite group

5.2.3.1 Roggianite I Ca^{2+}_{16} $(H_2O)_{16}$ I $[Al_{16}Si_{32}O_{88}(OH)_{16}]$ – **–RON**:
type material

Named after A.G. Roggiano, Italian mineral collector.

Remarks: Roggianite has a structure containing an interrupted framework of tetrahedra, it is accepted as zeolite because other zeolitic properties prevail.

Luster: silky, translucent
Channel system(s): [001] **12** 4.3 x 4.3*
Framework density: 18.2 T/ nm^3
Cages/cavities: none
Cleavage: none
Color: white, colorless
Crystallographic data: tetragonal, I4mcm
                          a = 1.833nm, c = 0.916nm

Hardness: 4
SBU(s): 4, combinations

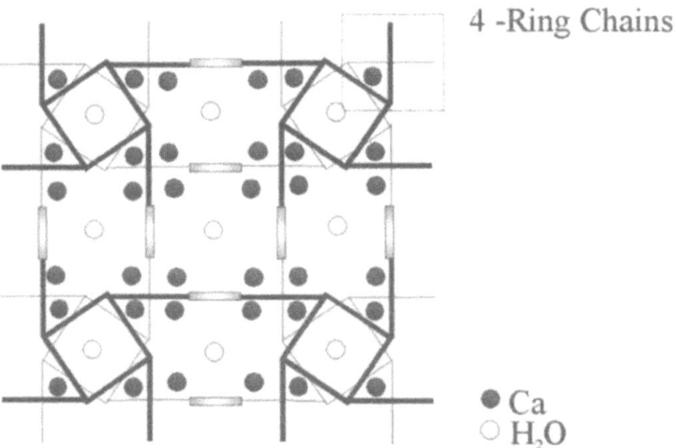

Figure 5.23: Projection of the Roggianite crystal structure in a modified representation [116,117].
Arrangement of the four rings c-axis, with view into the channel system.

Synthesis conditions:

High pressure hydrothermal treatment of synthetic glasses of Roggianite composition ($16CaO \times 8Al_2O_3 \times 32SiO_2$).
Hydrothermal parameters: 42 days synthesis time, 1000 bar water pressure, temperature interval of isothermal conditions: 80°C - 450°C. Optimum temperature of synthesis: 220°C.

Figure 5.24: Left: The SEM micrographs (3000 X) of the synthesized Roggianite. Right: Indexing of the Roggianite crystal revealing the tetragonal symmetry. The zeolite channels pass along c.

Other synthesis approaches: we are not aware of other synthesis approaches.

5.2 Zeolites with Singly Connected 4-Ring Chains

5.2.4 Yugawaralite group

5.2.4.1 Yugawaralite I Ca^{2+}_2 $(H_2O)_8$ I $[Al_4Si_{12}O_{32}]$ – YUG: type material

Named after the locality Yugawara hot springs, Honshu, Japan.

Remarks: piezoelectric
Luster: silky, translucent
Channel system(s): [100] **8** 2.8 x 3.6* ↔ [001] **8** 3.1 x 5.0*
Framework density: 18.3 T/ nm³
Cages/cavities: only given by channel intersections
Cleavage: along (011) incompletely, according (041) and (001) clearly
Color: white, colorless
Crystallographic data: monoclinic, Pc, a = 0.673nm, b = 1.395nm,
c = 1.003nm, β = 111.5°

Hardness: 5
SBU(s): 4

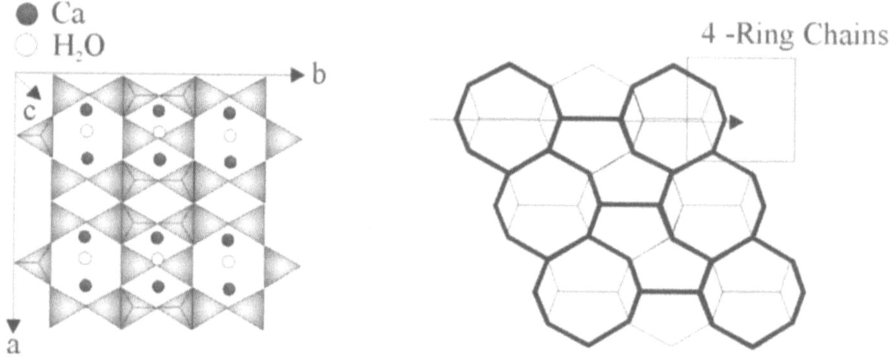

Figure 5.25: Projection of the Yugawaralite framework [118,119,120,121].
Right: Arrangement of the four ring chains.

Synthesis conditions:

High pressure hydrothermal treatment of synthetic water free glasses of Yugawaralite composition ($2CaO$ x $2Al_2O_3$ x $12SiO_2$).
Hydrothermal parameters: 42 days synthesis time, 1000 bar water pressure, temperature 230°C.

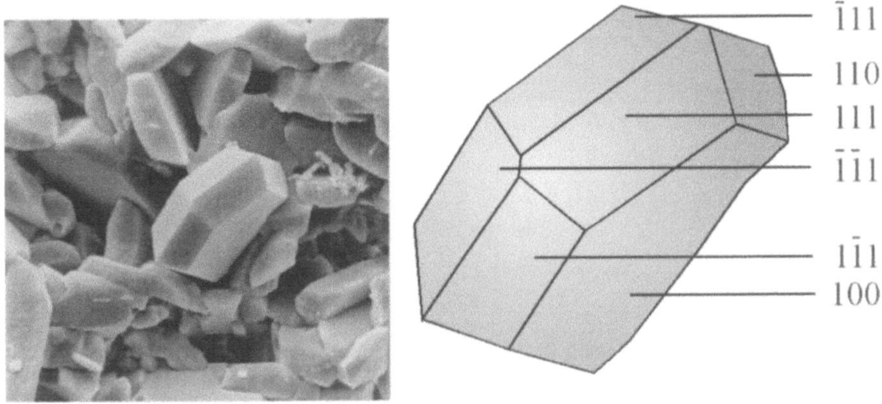

Figure 5.26: Left: The SEM micrographs (2000 X) of synthesized Yugawaralite Right: Indexing of the Yugawaralite crystal (monoclinic).

Other synthesis approaches: See [122,123,124,125].

5.3 Zeolites with Double Connected 4-Ring Chains

5.3.1 Gismondine group

5.3.1.1 Gismondine | Ca^{2+}_4 $(H_2O)_{16}$ | $[Al_8Si_8O_{32}]$ – GIS:

type material

Named after C.G. Gismondi, Italian mineralogist.

Remarks: Natural Gismondine is found to be monoclinic while our fresh synthesized material is tetragonal.

Luster: silky, translucent

Channel system(s): {[100] **8** 3.1 x 4.5 ↔ [010] **8** 2.8 x 4.8}***
Variable in size (considerable flexibility of framework)

Framework density: 15.3 T/ nm^3

Cages/cavities: only given by channel intersections

Cleavage: according ($\bar{2}$ 32)

Color: white, gray, colorless

Crystallographic data: monoclinic,   P2$_1$/c, a = 1.062nm, b = 1.062nm, c = 0.984nm, β = 92.416°

Hardness: 4.5

SBU(s): 4 and 8

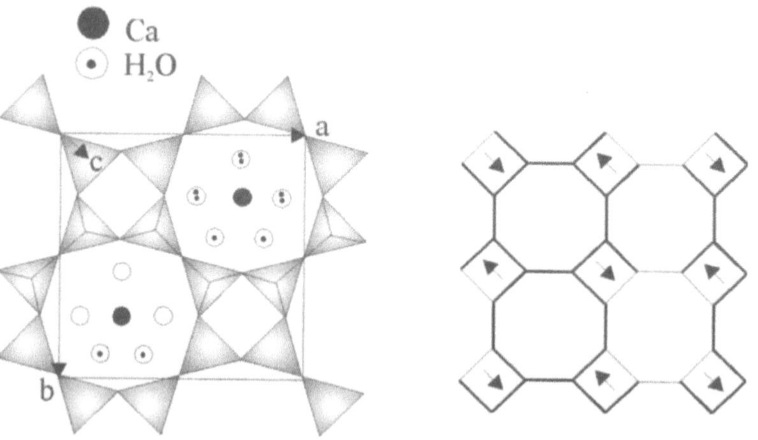

Fig. 5.27: Projection of the Gismondine structure [126,127,128,129].
Left : View into the channel system in c-direction with cations and water molecules. Right : The framework representation of Gismondine.

Synthesis conditions [130]:

High pressure hydrothermal treatment of synthetic water free glasses of Gismondine composition ($2CaO \times 2Al_2O_3 \times 4SiO_2$).
Hydrothermal parameters: 28 days synthesis time, 1000 bar water pressure, temperature interval 200°C to 250°C.

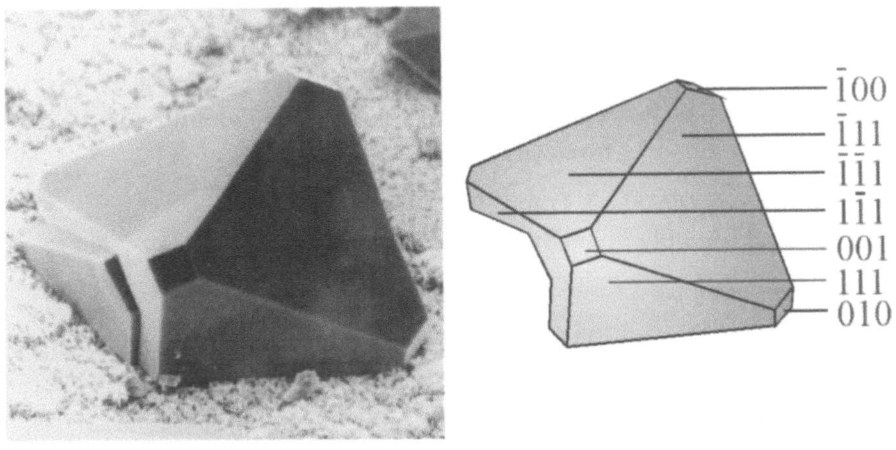

Figure 5.28: Left - the SEM micrographs (2000 X) of synthesized Gismondine. Right: Indexing of tetragonal Gismondine crystal.

Other synthesis approaches: See [131,132,133,134,52,135,136,137].

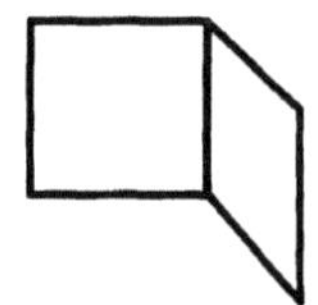

5.3 Zeolites with Double Connected 4-Ring Chains

5.3.1 Gismondine group

5.3.1.2 Garronite | $Na^+ Ca^{2+}_{2.5} (H_2O)_{13}$ | $[Al_6Si_{10}O_{32}]$ – GIS

Named after the locality Garron plateau, Northern Ireland.

Luster: translucent

Channel system(s): {[100] **8** 3.1 x 4.5 ↔ [010] **8** 2.8 x 4.8}***
                     Variable in size (considerable flexibility of framework)

Framework density: 15.3 T/ nm^3
Cages/cavities: only given by channel intersections

Cleavage: good, according to [100] and [010]
Color: colorless

Crystallographic data: tetragonal, $I\bar{4}m2$,
                         a = 0.985nm, c = 1.032nm

Hardness: 4.5
SBU(s): 4 and 8

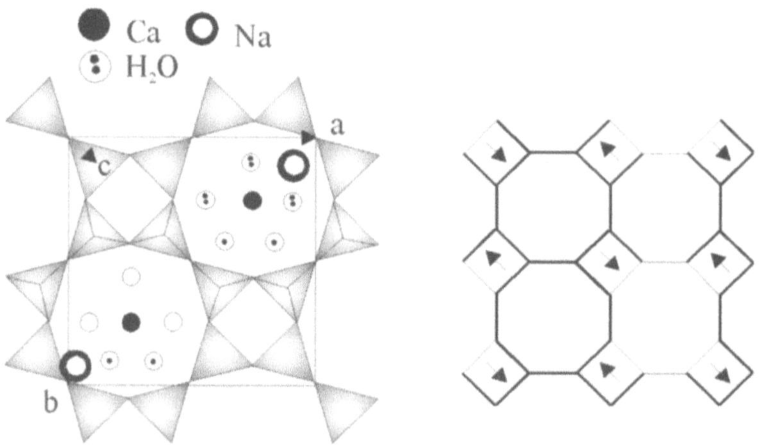

Fig. 5.29: Projection of the Garronite structure [138,139,140].
Left : View into the channel system of the crystal structure in c-direction with cations and water molecules. Right : The framework representation of Garronite.

Synthesis conditions [130]:

High pressure hydrothermal treatment of synthetic water free glasses of Garronite composition ($1Na_2O$ x $2.5CaO$ x $3Al_2O_3$ x $10SiO_2$).
Hydrothermal parameters: 28 days synthesis time, 1000 bar water pressure, temperature interval 200°C to 250°C.

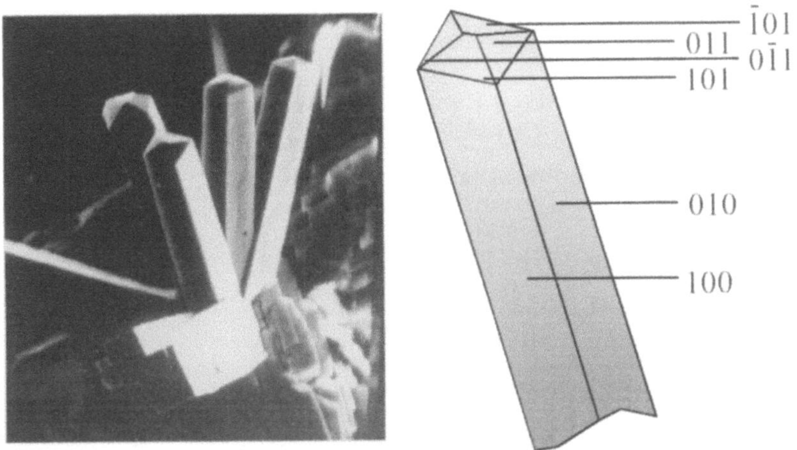

Figure 5.30: Left - the SEM micrographs (2000 X) of synthesized Garronite. Right: Indexing of tetragonal Garronite crystal.

Other synthesis approaches: See [132,141].

5.3 Zeolites with Double Connected 4-Ring Chains

5.3.1 Gismondine group

5.3.1.3 Amicite I K^+_2 Na^+_4 $(H_2O)_{10}$ I $[Al_8Si_8O_{32}]$ – GIS:

Named after G.B. Amici, inventor of the Amici lens.

Luster: translucent
Channel system(s): {[100] **8** 3.1 x 4.5 ↔ [010] **8** 2.8 x 4.8}***
             Variable in size (considerable flexibility of framework)

Framework density: 15.3 T/ nm^3
Cages/cavities: only given by channel intersections

Cleavage: none
Color: white, colorless
Crystallographic data: monoclinic, I2, a = 1.032nm,
             b = 1.042nm, c = 0.988nm, β = 88.316°

Hardness: 4.5
SBU(s): 4 and 8

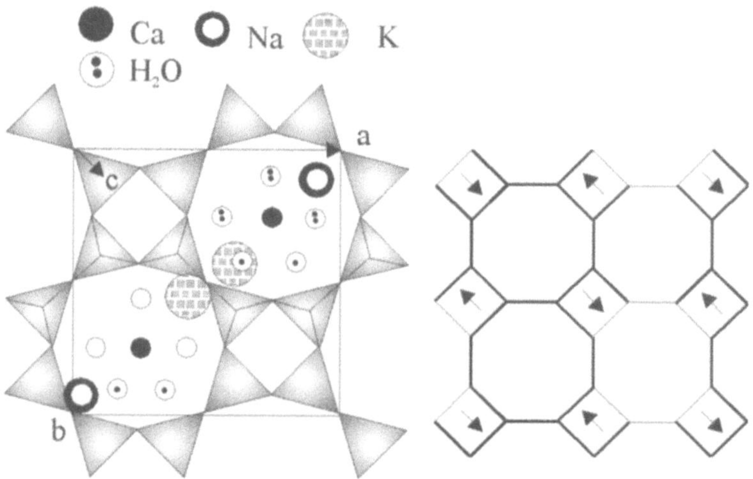

Fig. 5.31: Projection of the Amicite structure [142].
Left : View into the channel system in c-direction with cations and water molecules. Right : The framework representation of Amicite.

Synthesis conditions [130]:

High pressure hydrothermal treatment of synthetic water free glasses of Amicite composition ($1K_2O$ x $2Na_2O$ x $4Al_2O_3$ x $8SiO_2$).
Hydrothermal parameters: 28 days synthesis time, 1000 bar water pressure, temperature interval 200°C to 250°C.

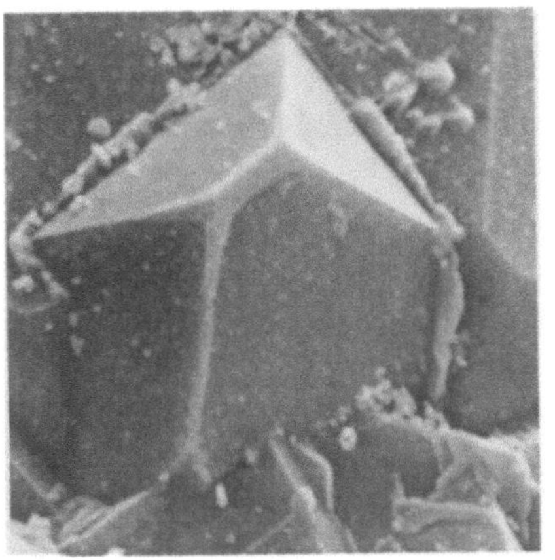
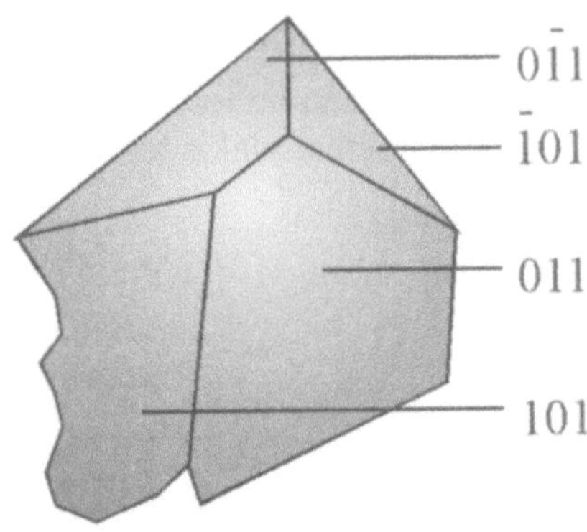

Figure 5.32: Left - the SEM micrographs (300 X) of synthesized Amicite. Right: Indexing of monoclinic Amicite crystal.

Other synthesis approaches: See [132].

5.3 Zeolites with Double Connected 4-Ring Chains

5.3.1 Gismondine group

5.3.1.4 Gobbinsite I Na^{+}_{5} $(H_2O)_{11}$ **I** $[Al_5Si_{11}O_{32}]$ – **GIS:**

Named after the locality Gobbins area, Co. Antrim, Northern Ireland.

Remarks: Natural Gobbinsite is found to be orthorhombic while our fresh synthesized material is tetragonal.

Luster: glassy
Channel system(s): {[100] **8** 3.1 x 4.5 ↔ [010] **8** 2.8 x 4.8}***
Variable in size (considerable flexibility of framework)

Framework density: 15.3 T/ nm^3
Cages/cavities: only given by channel intersections

Cleavage: none
Color: white
Crystallographic data: orthorhombic, $Pnm2_1$, a = 0.980nm,
b = 1.015nm, c = 1.010nm

Hardness: 4-5
SBU(s): 4 and 8

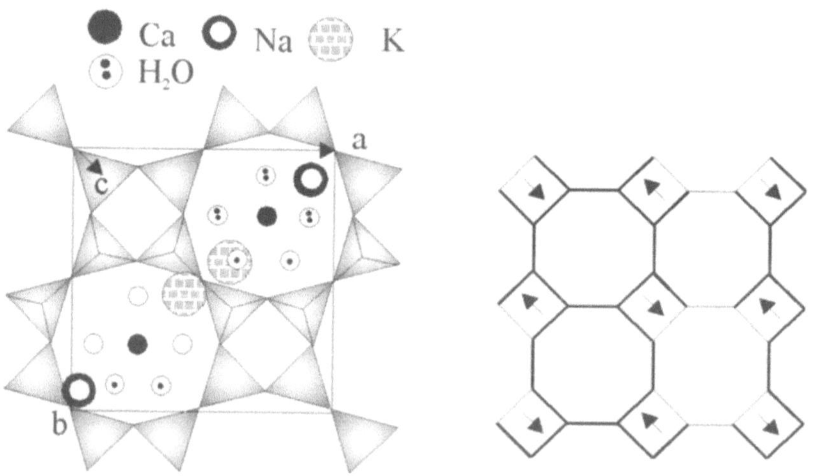

Fig. 5.33: Projection of the Gobbinsite structure [143,144].
Left : View into the channel system in c-direction with cations and water molecules. Right : The framework representation of Gobbinsite.

Synthesis conditions [130]:

High pressure hydrothermal treatment of synthetic water free glasses of Gobbinsite composition ($5Na_2O$ x $5Al_2O_3$ x $11SiO_2$).
Hydrothermal parameters: 28 days synthesis time, 1000 bar water pressure, temperature interval 200°C to 250°C.

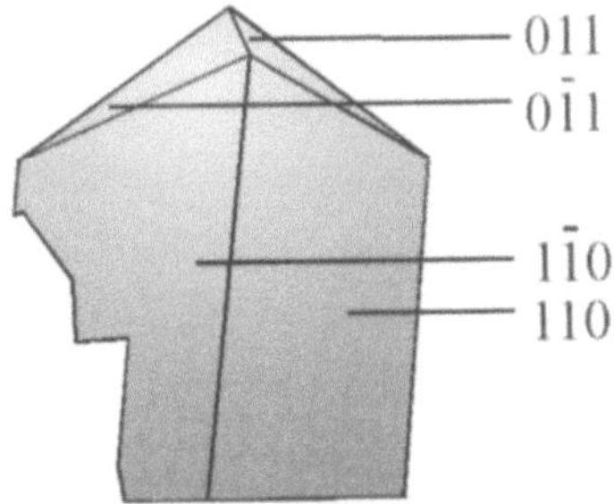

Figure 5.34: Left - the SEM micrographs (ca. 2000 X) of synthesized Gobbinsite. Right: Indexing of tetragonal Gobbinsite crystal.

Other synthesis approaches: See [132,145].

5.3 Zeolites with Double Connected 4-Ring Chains

5.3.2 Phillipsite group

5.3.2.1 Phillipsite | $K^+_2 Na^+ Ca^{2+}_{0.5} (H_2O)_{12}$ | $[Al_6Si_{10}O_{32}]$ – **PHI**:

type material

Named after W. Phillips, English mineralogist and founder of the Geological Society of London.

Luster: milky, translucent
Channel system(s): [100] **8** 3.8 x 3.8* ↔ [010] **8** 3.0 x 4.3*
                    ↔ [010] **8** 3.0 x 4.3*
Framework density: 15.8 T/ nm^3
Cages/cavities: only given by channel intersections
Cleavage: perfect according (100) and (010)
Color: white, yellowish
Crystallographic data: monoclinic, pseudo-orthorhombic,
                    $P2_1/m$, a = 0.988nm, b = 1.430nm, c = 1.434nm,
                    β = 89.466°

Hardness: 4 to 5
SBU(s): 4 and 8

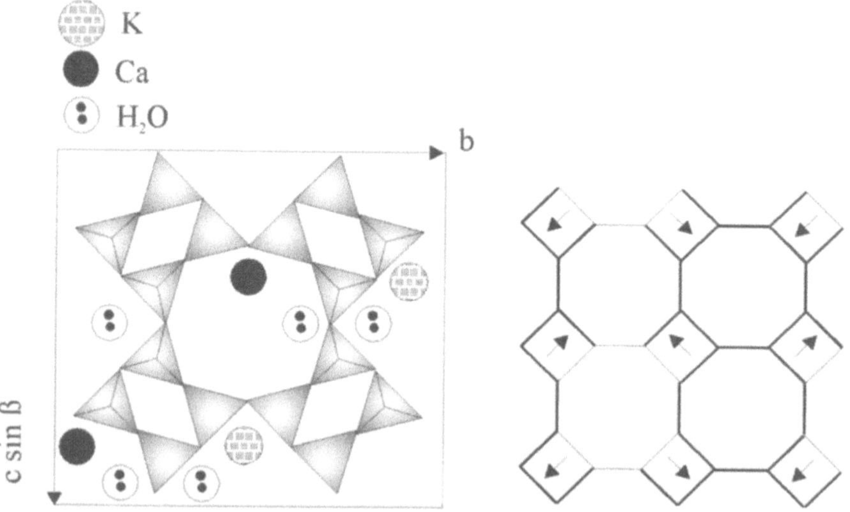

Fig. 5.35: Projection of the Phillipsite structure [146,147,148,149].
Left : View into the channel system in a-direction with cations and water molecules. Right : The framework representation of Phillipsite.

Synthesis conditions:

High pressure hydrothermal treatment of synthetic water free glasses of Phillipsite composition ($1K_2O$ x $0.5Na_2O$ x $0.5CaO$ x $3Al_2O_3$ x $10SiO_2$). Hydrothermal parameters: 60 days synthesis time, 1000 bar water pressure, temperature 200°C to 250°C.

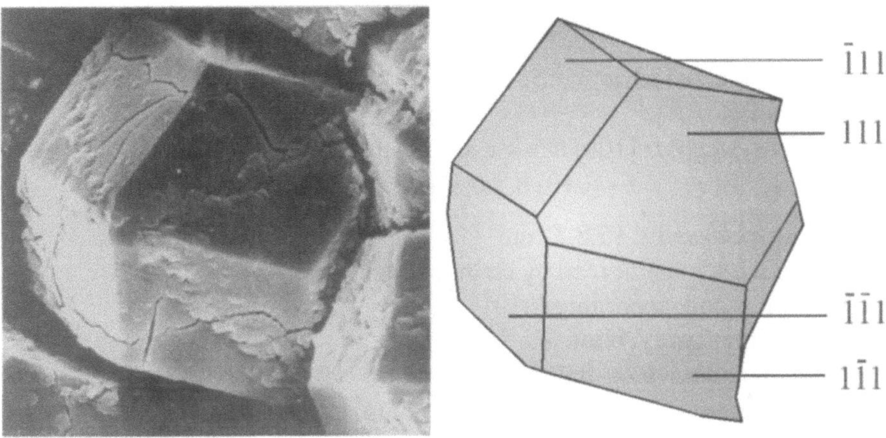

Figure 5.36: Left - the SEM micrographs (2000 X) of synthesized Phillipsite. Right: Indexing of the monoclinic Phillipsite crystal.

Other synthesis approaches: See [150,151,152,141,52,153,154,155,156,157,158,159,160,161].

5.3 Zeolites with Double Connected 4-Ring Chains

5.3.2 Phillipsite group

5.3.2.2 Harmotome I $Ba^{2+}_2 Ca^{2+}_{0.5} Na^+ (H_2O)_{12}$ I $[Al_5Si_{11}O_{32}]$ – PHI:

Name from the Greek: αρμοσ = joint and τομη = cut, according to the tendency to cleave along junctions (twin planes).

Remarks: piezoelectric
Luster: silky
Channel system(s): [100] **8** 3.8 x 3.8* ↔ [010] **8** 3.0 x 4.3*
           ↔ [010] **8** 3.0 x 4.3*

Framework density: 15.8 T/ nm^3
Cages/cavities: only given by channel intersections
Cleavage: good, according (010)
Color: white, gray, yellowish
Crystallographic data: monoclinic, pseudo-orthorhombic,
        $P2_1/m$, a = 0.988nm, b = 1.414nm, c = 1.427nm,
        β = 90.166°

Hardness: 4.5
SBU(s): 4 and 8

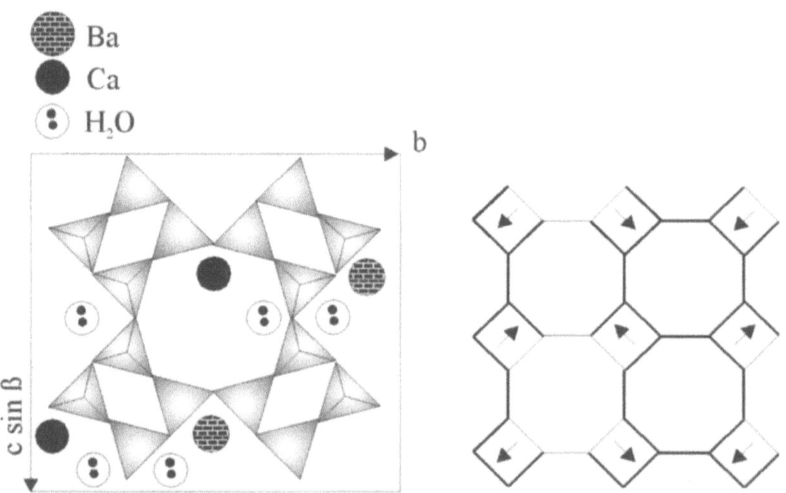

Fig. 5.37: Projection of the Harmotome structure [162,163,164].
Left : View into the channel system in c-direction with cations and water molecules. Right : The framework representation of Harmotome.

Synthesis conditions:

High pressure hydrothermal treatment of synthetic water free glasses of Harmotome composition (2BaCl$_2$ x 0.5CaO 0.5Na$_2$O x 2.5Al$_2$O$_3$ x 11SiO$_2$).
Hydrothermal parameters: 60 days synthesis time, 1000 bar water pressure, temperature 200°C to 250°C.

Figure 5.38: Left - the SEM micrographs (1000 X) of synthesized Harmotome. Right: Indexing of a twinned monoclinic Harmotome crystal.

Other synthesis approaches: See [165,166,167,168,169,170,171].

5.3 Zeolites with Double Connected 4-Ring Chains

5.3.3 Merlinoite group

5.3.3.1 Merlinoite I $(Na^+ K^+)_5 (Ca^{2+} Ba^{2+})_2 (H_2O)_{24}$ I $[Al_9Si_{23}O_{64}]$ – **MER**:
type material

Named after S. Merlino, Italian Crystallographer, University of Pisa

Luster: silky
Channel system(s): [100] **8** 3.1 x 3.5* ↔ [010] **8** 2.7 x 3.6*
↔ [001] {**8** 3.4 x 5.1* + **8** 3.3 x 3.3*}
Framework density: 16.0 T/ nm³
Cages/cavities: ε - cages, see below
Cleavage: none
Color: white, gray, colorless
Crystallographic data: orthorhombic, Immm, a = 1.412nm, b = 1.423nm,
c = 0.995nm

Hardness: 4-5
SBU(s): 4, 8 and 8-8

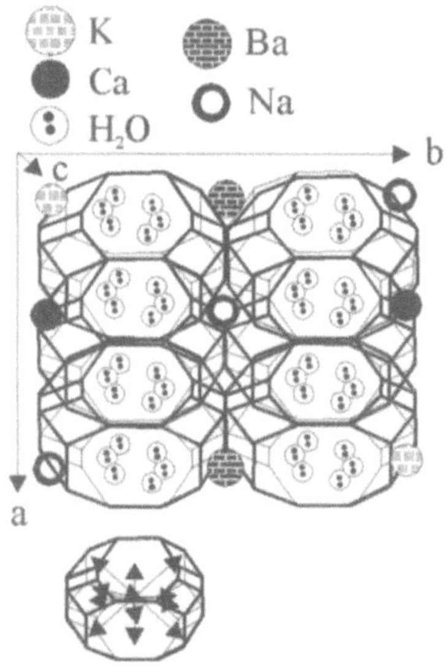

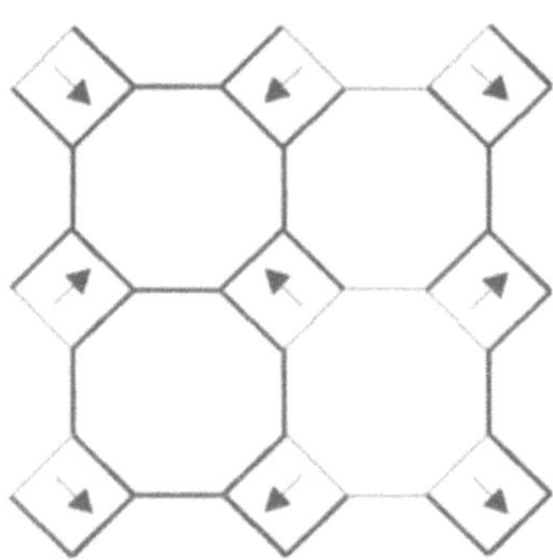

Fig. 5.39: Projection of the Merlinoite structure [172,173,174].
Left : View into the channel (above) and cage/cavity system (below) in c-direction with cations and water molecules. Right : The framework representation of Merlinoite, closly related to Phillipsite.

Synthesis conditions:

High pressure hydrothermal treatment of synthetic water free glasses obtained by co-melting of $BaCO_3$, $CaCO_3$, $NaOH$, KOH, Al_2O_3, SiO_2 in the appropriate proportions given above.
Hydrothermal parameters: 60 days synthesis time, 1000 bar water pressure, temperature interval 200°C to 250°C.

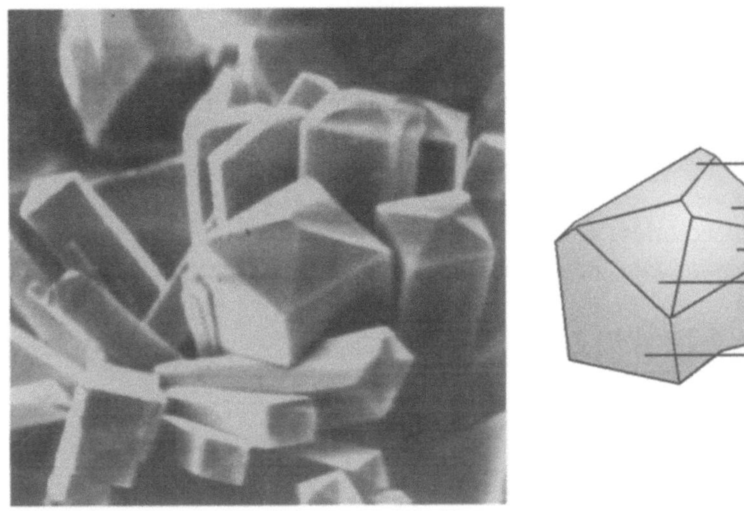

Figure 5.40: Left - the SEM micrographs (1500 X) of synthesized Merlinoite. Right: Indexing of orthorhombic Merlinoite crystal.

Other synthesis approaches: See [52,145,175,176,177,178,179,180,181,182].

5.3 Zeolites with Double Connected 4-Ring Chains

5.3.4 Mazzite group

5.3.4.1 Mazzite I K^+_3 $Ca^{2+}_{1.5}$ Mg^{2+}_2 $(H_2O)_{28}$ I $[Al_{10}Si_{26}O_{72}]$ – **MAZ**: type material

Named after F. Mazzi, Italian Mineralogist, University of Pisa

Luster: translucent
Channel system(s): [0001] **12** 7.4 x 7.4* I [0001] **8** 3.1 x 3.1***
Framework density: 16.1 T/ nm^3
Cages/cavities: Mazzite cavities (see below)
Cleavage: none
Color: white, colorless
Crystallographic data: hexagonal, P6$_3$/mmc, a = 1.839nm, c = 0.765nm
Hardness: 4
SBU(s): 5-1, 4-2 and 4

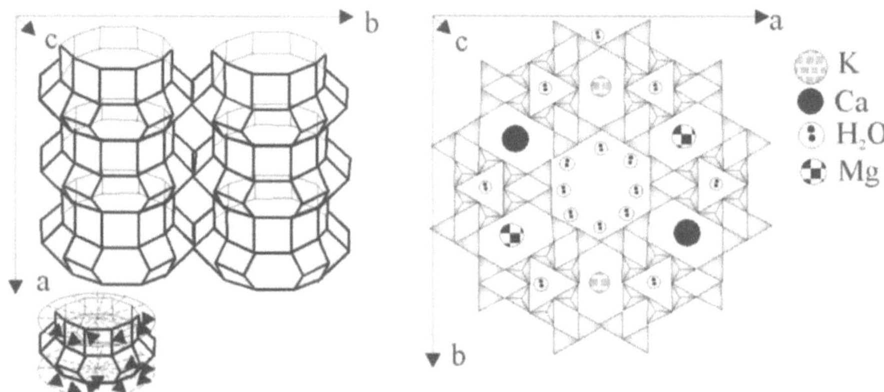

Fig. 5.41: Projection of the Mazzite structure [183,184,185].
Left : Channel system in c-direction and void configuration (below). Right : The framework representation of Mazzite with cations and water molecules.

Synthesis conditions [186]:

High pressure hydrothermal treatment of synthetic water free glasses corresponding the Mazzite composition (3KOH x 1.5CaO x 2MgO x 5Al$_2$O$_3$ x 26SiO$_2$).
Hydrothermal parameters: 60 days synthesis time, 1000 bar water pressure, temperature interval 150°C -250°C.

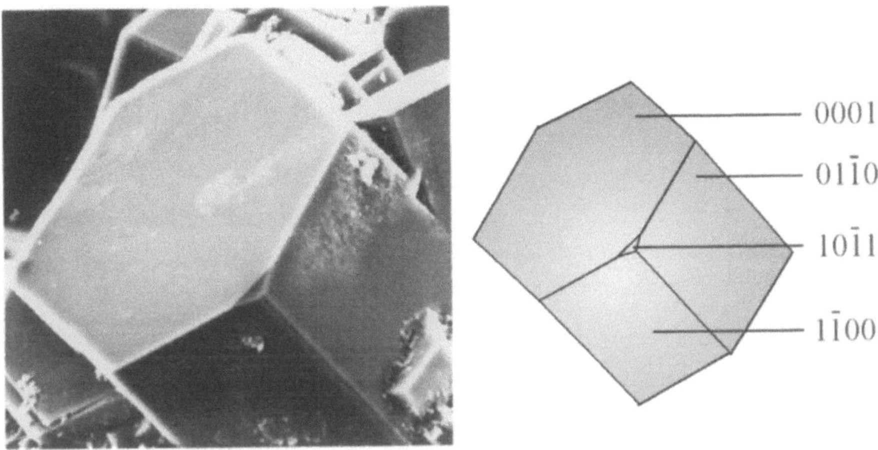

Figure 5.42: Left - the SEM micrographs (700 X) of synthesized Mazzite. Right: Indexing of hexagonal Mazzite crystal.

Other synthesis approaches: See [187,188,189,190,191,192,193,194,195].

5.3 Zeolites with Double Connected 4-Ring Chains

5.3.5 Paulingite group

5.3.5.1 Paulingite | K^+_{68} Na^+_{12} Ca^{2+}_{41} $(H_2O)_{705}$ | $[Al_{162}Si_{500}O_{1344}]$ – **PAU**:
type material

Named after L.C. Pauling, American Chemist and Nobel-Prize winner, California Institute of Technology.

Luster: glassy
Channel system(s): <100> **8** 3.6 x 3.6*** | <100> **8** 3.6 x 3.6***
Framework density: 15.5 T/ nm^3
Cages/cavities: double twelve rings
Cleavage: none
Color: colorless, white, yellowish to orange
Crystallographic data: cubic, Im3m, a = 3.509nm
Hardness: 5
SBU(s): 4

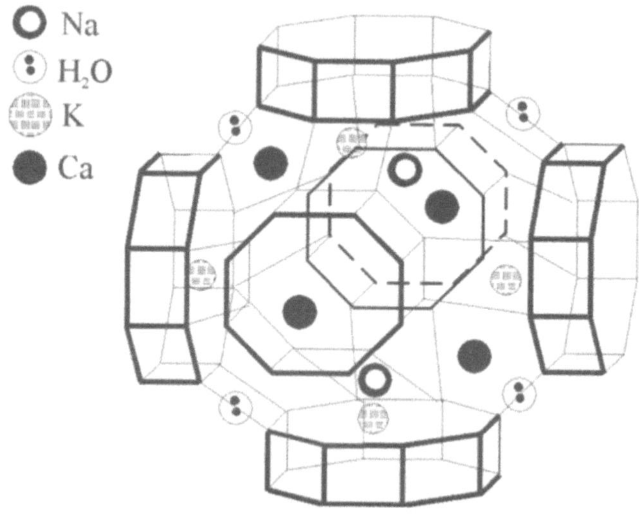

Fig. 5.43: Projection of the Paulingite structure [196,197].
Left : View into the framework with the void system including cations and water molecules.

Synthesis conditions:

High pressure hydrothermal treatment of synthetic water free glasses obtained by co-melting of KOH, NaOH, $CaCO_3$, Al_2O_3 and SiO_2, corresponding the Paulingite composition
Hydrothermal parameters: 60 days synthesis time, 1000 bar water pressure, temperature 200°C -250°C.

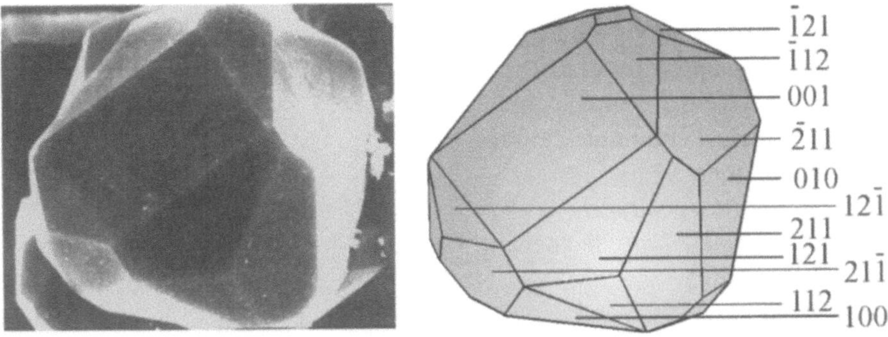

Figure 5.44: Left - the SEM micrographs (2000 X) of synthesized Paulingite. Right: Indexing of the Paulingite crystal.

Other synthesis approaches: See [198,199,200].

5.4 Zeolites with 6-Rings

5.4.1 Gmelinite group

5.4.1.1 Gmelinite | Na$^+_8$ (H$_2$O)$_{22}$ | [Al$_8$Si$_{16}$O$_{48}$] – GME: type material

Named after G.C. Gmelin, German Chemist, University of Tübingen.

Luster: translucent
Channel system(s): [0001] **12** 7.0 x 7.0* $\leftrightarrow$ $\perp$ [0001] **8** 3.6 x 3.9**
Framework density: 14.6 T/ nm^3
Cages/cavities: Gmelinite cavities
Cleavage: none
Color: colorless, colored
Crystallographic data: hexagonal, P6$_3$/mmc, a = 1.375nm, c = 1.006nm
Hardness: 4 to 5
SBU(s): 6, 6-6, 4, 4-2 and 8

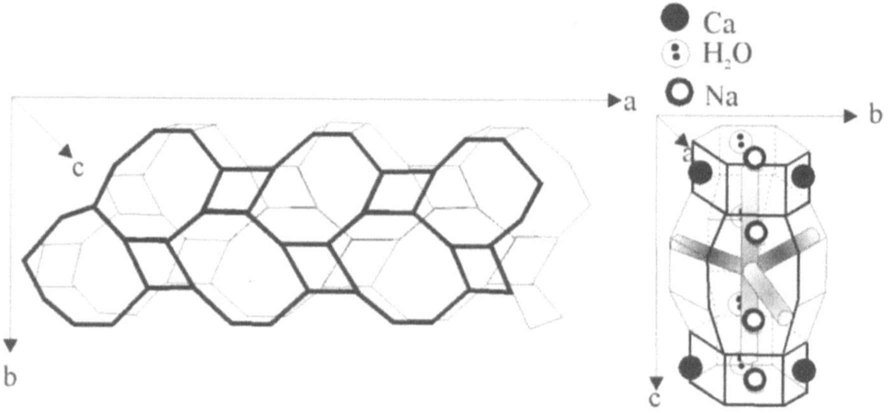

Fig. 5.45: Projection of the Gmelinite structure [201,202,203,204,205,206,207].
Left : Tubular building of the Gmelinite framework. Right : Structure and channel system with cations and water molecules.

Synthesis conditions [208]:

High pressure hydrothermal treatment of synthetic water free glasses of Gmelinite composition ($1Na_2O$ x $0.6(K_2Ca)O$ x $1Al_2O_3$ x $4SiO_2$).
Hydrothermal parameters: 60 days synthesis time, 1000 bar water pressure, temperature interval 170°C to 270°C.

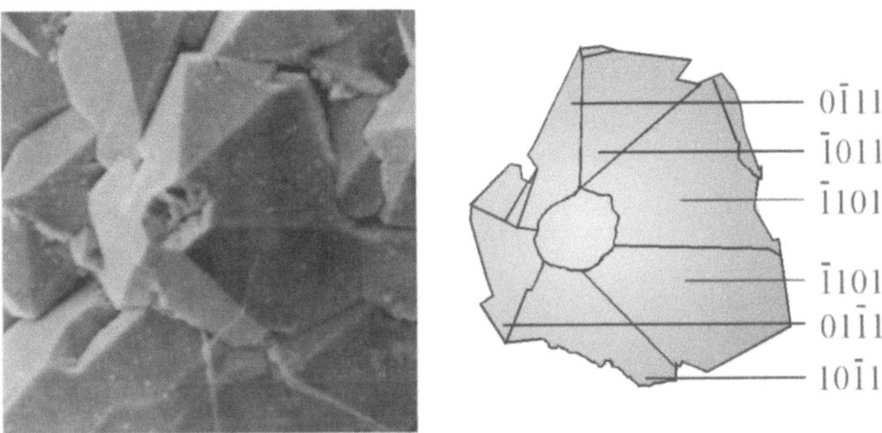

Figure 5.46: Left - the SEM micrographs (1000 X) of synthesized Gmelinite. Right: Indexing of a hexagonal Gmelinite crystal.

Other synthesis approaches: See [131,209,210,211,212,213,214,215,216,217,218].

5.4 Zeolites with 6-Rings

5.4.2 Chabazite group

5.4.2.1 Chabazite I Ca^{2+}_2 $(H_2O)_{12}$ I $[Al_4Si_8O_{24}]$ – **CHA**: type material

From the Greek word χαβαζιον - the name of an unknown substance in the myth of Orpheus.

Luster: glassy
Channel system(s): ⊥ [0001] **8** 3.8 x 3.8***
Variable in size (considerable flexibility of framework)

Framework density: 14.5 T/ nm^3
Cages/cavities: Chabazite cavities

Cleavage: according ($10\overline{1}1$)
Color: white, reddish, brownish

Crystallographic data: trigonal rhomboedric, R$\overline{3}$m, a = 1.37nm,
c = 15.02nm

Hardness: 4 to 5
SBU(s): 6, 6-6 and 4

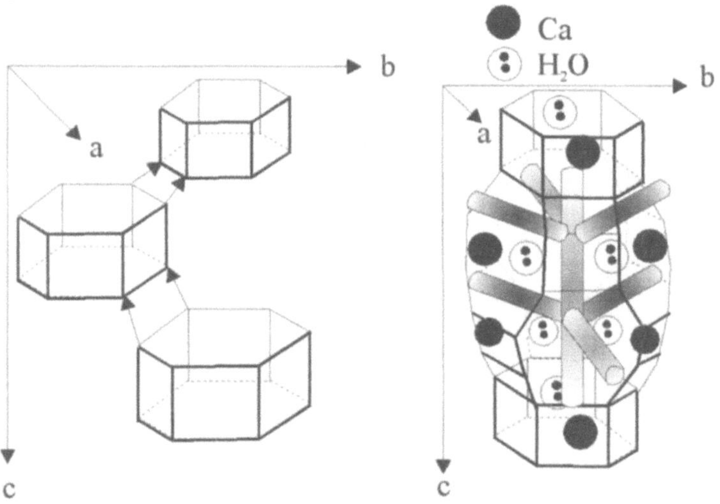

Fig. 5.47: Projection of the Chabazite structure [219,220,221,222,223,224,225,226,227].
Left : The framework of Chabazite. Right: The Chabazite void system with cations and water molecules. Note the location of the channel system.

Synthesis conditions [208]:

High pressure hydrothermal treatment of synthetic water free glasses of Chabazite composition (1CaO x 0.6(K,Li)$_2$O x 1Al$_2$O$_3$ x 4SiO$_2$).
Hydrothermal parameters: 60 days synthesis time, 1000 bar water pressure, temperature interval 170°C to 270°C.

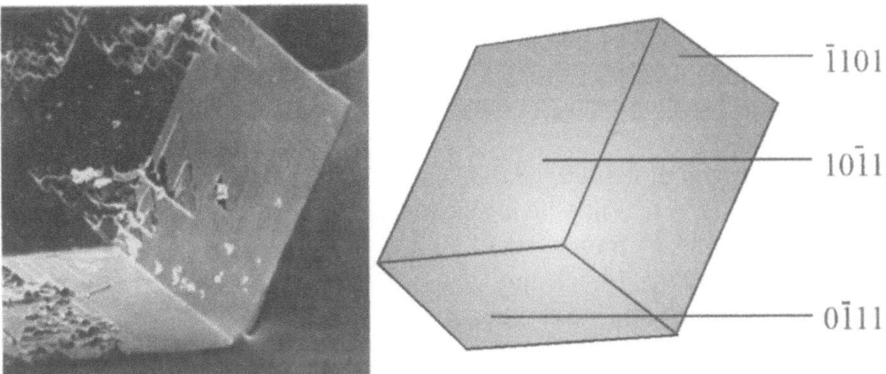

Figure 5.48: Left - the SEM micrographs (2000 X) of synthesized Chabazite. Right: Indexing of a trigonal Chabazite crystal.

Other synthesis approaches: See [228,229,230,231,192,232,233,234,235,236,237,238,239].

5.4 Zeolites with 6-Rings

5.4.2 Chabazite group

5.4.2.2 Willhendersonite | $K^+_2 Ca^{2+}_2 (H_2O)_{10}$ | $[Al_6Si_6O_{24}]$ – CHA

Named after Dr. William A. Henderson Jr., American Mineralogist.

Remarks: Herschelite is discredited as a zeolite species: although developing its proper morphology during high pressure hydrothermal synthesis, it has the Chabazite framework.

Luster: glassy

Channel system(s): ⊥ [0001] **8** 3.8 x 3.8***

Variable in size (considerable flexibility of framework)

Framework density: 14.5 T/ nm^3

Cages/cavities: Chabazite cavities

Cleavage: according (0001), (10$\bar{1}$0) and ($\bar{1}$100)

Color: colorless

Crystallographic data: triclinic, P$\bar{1}$,

a = 0.92nm, b = 0.918nm, c = 0.949nm,
α = 92.6°, β = 92.43°, γ = 90.05°

Hardness: 4 to 5

SBU(s): 6, 6-6 and 4

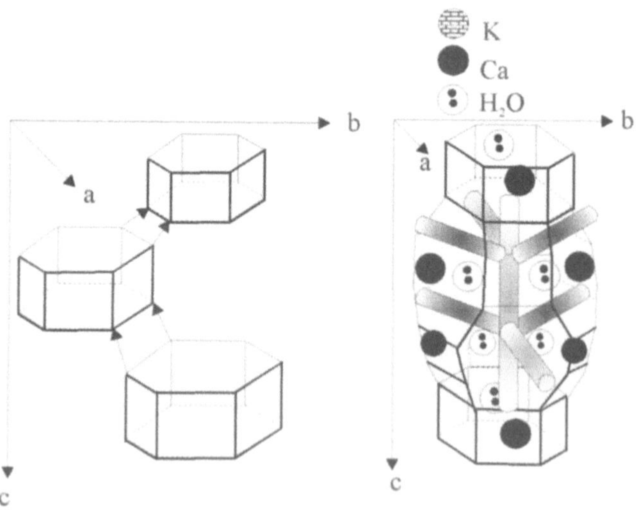

Fig. 5.49: Projection of the wilhendersonite structure [240,241].
Left : The framework of wilhendersonite. Right : Channel and void system with cations and water molecules.

Synthesis conditions:

High pressure hydrothermal treatment of synthetic water free glasses of Willhendersonite composition (ca. $1CaO \times 0.6Li_2O \times 1.5K_2O \times 1Al_2O_3 \times 4SiO_2$).

Hydrothermal parameters: 60 days synthesis time, 1000 bar water pressure, temperature interval 170°C to 270°C.

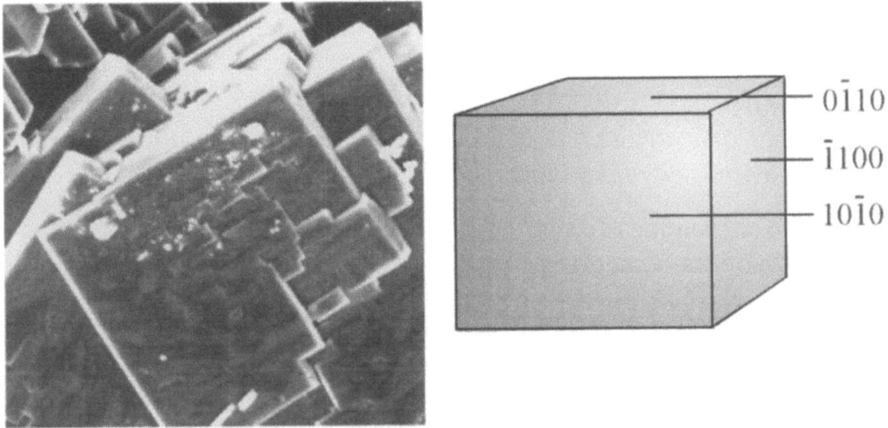

Figure 5.50: Left - the SEM micrographs (1000 X) of synthesized Willhendersonite. Right: Indexing of triclinic Willhendersonite crystal.

The authors are not aware of any other Willhendersonite syntheses.

5.4 Zeolites with 6-Rings

5.4.3 Levyne group

5.4.3.1 Levyne | Na$^+$ Ca$^{2+}_{2.5}$ (H$_2$O)$_{18}$ | [Al$_6$Si$_{12}$O$_{36}$] – LEV:

type material

Named after A. Levy, French mathematician and crystallographer, University of Paris

Luster: glassy
Channel system(s): ⊥ [0001] **8** 3.6 x 4.8**
Framework density: 15.2 T/ nm^3
Cages/cavities: Levyne cavities

Cleavage: according (10$\bar{1}$1)
Color: colorless, sometimes yellowish

Crystallographic data: trigonal, R$\bar{3}$m, a = 1.335nm, c = 2.290nm
Hardness: 4 to 4.5
SBU(s): 6

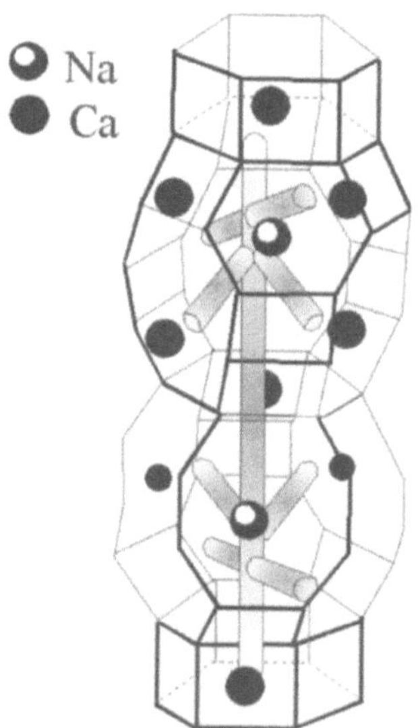

Fig. 5.51: Projection of the Levyne structure [242,243].
Structure, channel and void system of Levyne as well as the positions of the cations.

Synthesis conditions [208]:

High pressure hydrothermal treatment of synthetic water free glasses of Levyne composition (5CaO x 1Na$_2$O x 6Al$_2$O$_3$ x 24SiO$_2$).
Hydrothermal parameters: 60 days synthesis time, 1000 bar water pressure, temperature interval 170°C to 270°C.

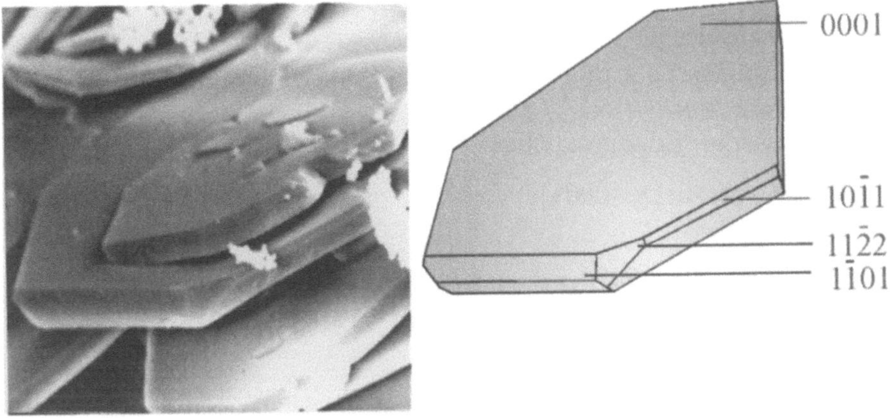

Figure 5.52: Left - the SEM micrographs (1000 X) of synthesized Levyne. Right: Indexing of trigonal Levyne crystal.

Other synthesis approaches: See [244,245,246,52,247].

5.4 Zeolites with 6-Rings

5.4.4 Erionite group

5.4.4.1 Erionite I Na$^+$ K$^+_2$ Mg^{2+} Ca$^{2+}_{1.5}$ (H$_2$O)$_{28}$ I [Al$_8$Si$_{28}$O$_{72}$]

– **ERI**: type material

Name from the Greek εριον = "wool", due to its appearance.

Luster: translucent
Channel system(s): ⊥ [0001] **8** 3.6 x 5.1**
Framework density: 15.7 T/ nm^3
Cages/cavities: Erionite cavities

Cleavage: parallel to [0001]
Color: colorless

Crystallographic data: hexagonal, P$_3$/mmc, a = 1.315nm,
                                                c = 1.505nm

Hardness: 4
SBU(s): 6 and 4

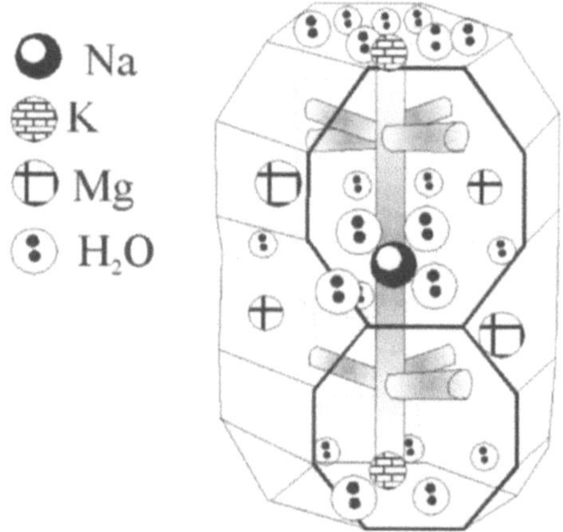

Fig. 5.53: Projection of the Erionite structure [248,249,250,251,252].
Structure, channel and void system of Erionite as well as the positions of the cations.

Synthesis conditions [208]:

High pressure hydrothermal treatment of synthetic water free glasses corresponding the Erionite composition ($1Na_2O$ x $2K_2O$ x $2MgO$ x $3CaO$ x $8Al_2O_3$ x $54SiO_2$ x $3.5Fe_2O_3$: iron oxide serves as structure directing agent).
Hydrothermal parameters: 60 days synthesis time, 1000 bar water pressure, temperature interval 170°C to 270°C.

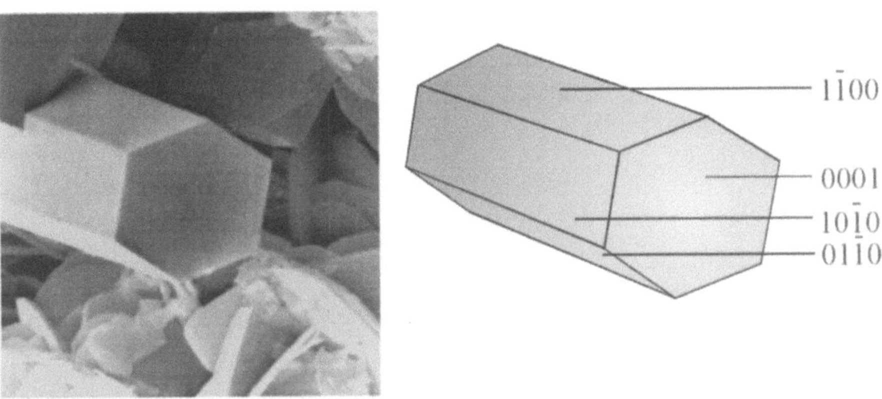

Figure 5.54: Left - the SEM micrographs (2000 X) of synthesized Erionite. Right: Indexing of the Erionite crystal (hexagonal).

Other synthesis approaches:

See [141,189,253,254,255,256,257,258,259,260,261,262,263,264,265].

5.4 Zeolites with 6-Rings

5.4.5 Offretite group

5.4.5.1 Offretite I $K^+ Mg^{2+} Ca^{2+} (H_2O)_{15}$ I $[Al_5Si_{13}O_{36}]$

– **OFF**: type material

Named afterA.J.J. Offret, French Mineralogist, University of Lyon.

Luster: translucent
Channel system(s): [0001] **12** 6.7 x 6.8* ↔ ⊥ [0001] **8** 3.6 x 4.9**
Framework density: 15.5 T/ nm^3
Cages/cavities: Offretite cavities
Cleavage: parallel [0001]
Color: colorless

Crystallographic data: hexagonal, $P\bar{6}m2$, a = 1.329nm,
c = 0.758nm

Hardness: 4
SBU(s): 6

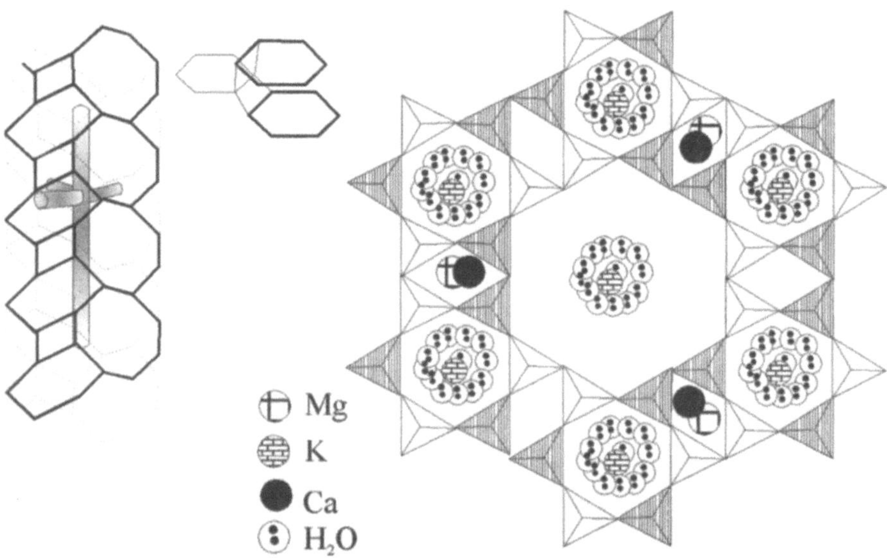

Fig. 5.55: Projection of the Offretite structure [266,267,268,269,270].
Left: Structure, and channel/cavity system of Offretite as well as interconnection
of the SBUs. Right: View of the Offretite structure from c-direction with cations
and water positions.

Synthesis conditions [208,271]:

High pressure hydrothermal treatment of synthetic water free glasses corresponding the Offretite composition ($1K_2O$ x $2MgO$ x $2CaO$ x $5Al_2O_3$ x $26SiO_2$ x $1.8Fe_2O_3$: iron oxide serves as structure directing agent). Hydrothermal parameters: 60 days synthesis time, 1000 bar water pressure, temperature interval 170°C to 270°C.

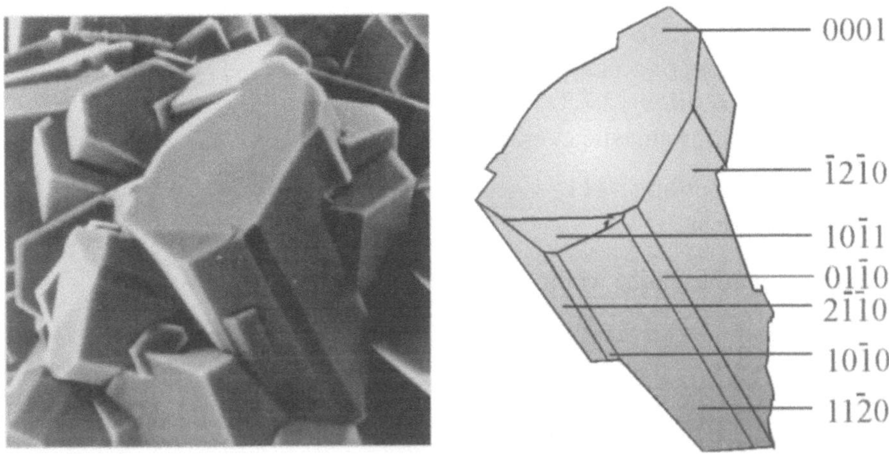

Figure 5.56: Left - the SEM micrographs (2000 X) of synthesized Offretite. Right: Indexing of the hexagonal Offretite crystal.

Other synthesis approaches:

See [253,255,256,260,261,262,272,273,274,275,276,277,278,279,280,281,282,283,284,285,286,287,288,289,290].

5.4 Zeolites with 6-Rings

5.4.6 Faujasite group

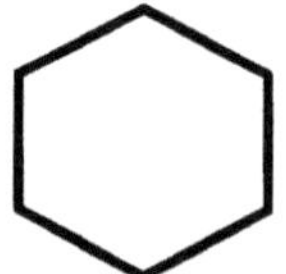

5.4.6.1 Faujasite | $Na^+_{20} Mg^{2+}_8 Ca^{2+}_{12} (H_2O)_{235}$ | $[Al_{60}Si_{132}O_{384}]$ – **FAU**:
type material

Named after B. Faujas de Saint Fond, French Naturalist.

Luster: glassy
Channel system(s): <111> **12** 7.4 x 7.4***
Framework density: 12.7 T/ nm^3
Cages/cavities: Faujasite cavities and Sodalite cages

Cleavage: according <111>
Color: colorless

Crystallographic data: cubic, $Fd\bar{3}m$, a = 2.46nm,

Hardness: 5
SBU(s): 6-6, 6-2, 6 and 4

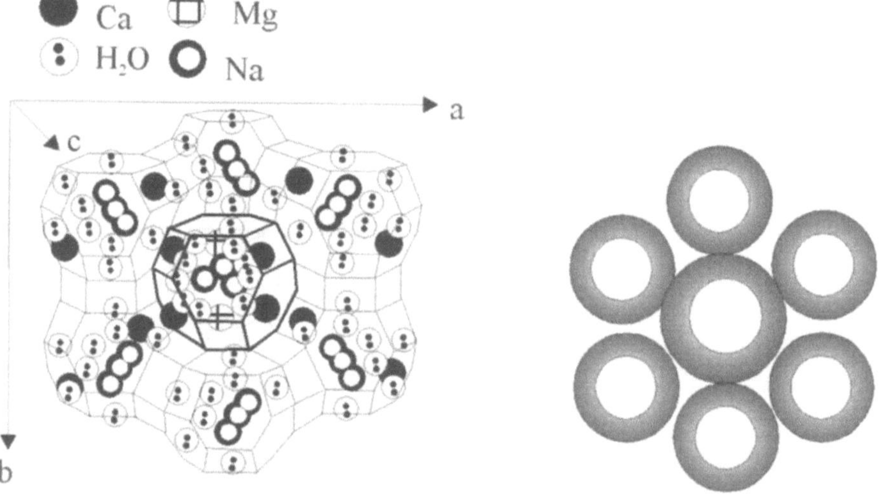

Fig. 5.57: Projection of the Faujasite structure [291,292,293,294,295,296,297,298].
Left : The Faujasite framework with the Sodalite cage arrangement and the central
Faujasite supercavity viewed from c with the cation and water positions. Right:
Model of the channel system viewed along c.

Synthesis conditions [208]:

High pressure hydrothermal treatment of synthetic water free glasses of the following composition: ($5Na_2O$ x $4MgO$ x $6CaO$ x $15Al_2O_3$ x $66SiO_2$ x $9.6(Sr,Ba,K_2)O$ x $4.8Fe_2O_3$: iron oxide serves as structure directing agent). Hydrothermal parameters: 60 days synthesis time, 1000 bar water pressure, temperature interval 170°C to 270°C.

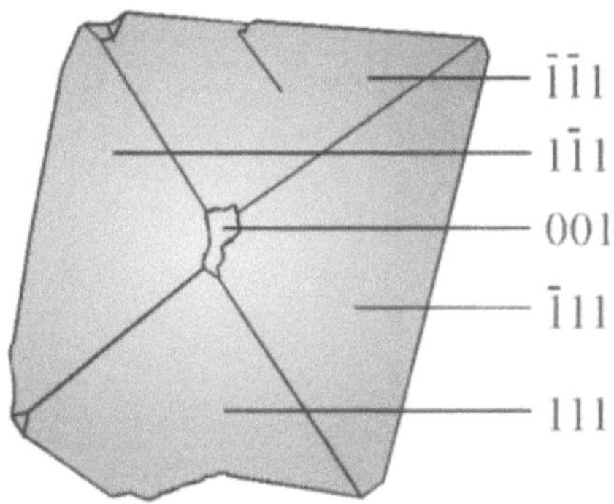

Figure 5.58: Left - SEM micrograph (2000 X) of synthesized Faujasite. Right: Indexing of cubic Faujasite crystal.

Other synthesis approaches: See [187,282,299,300,301,302,303,304,305,306,307,308,309,310, 311,312,313,314,315,316,317,318,319,320,321,322,323,324,325,326,327,328,329,330,331,332,333,334, 335,336,337,338].

5.4 Zeolites with 6-Rings

5.4.7 Goosecreekite group

5.4.7.1 Goosecreekite I $Ca^{2+}_2 (H_2O)_{10}$ I $[Al_4Si_{12}O_{32}]$ – **GOO**:

type material

Named after the locality Goose Creek quarry, Virginia, USA.

Luster: translucent
Channel system(s): [100] **8** 2.8 x 4.0* ↔ [010] **8** 2.7 x 4.1*
                         ↔ [001] **8** 2.9 x 4.7*
Framework density: 17.6 T/ nm^3
Cages/cavities: only given by channel intersections
Cleavage: according (010)
Color: white,
Crystallographic data: monoclinic, $P2_1/m$, a = 0.752nm, b = 1.756nm,
                         c = 0.735nm, β = 105.716°

Hardness: 4
SBU(s): 6-2

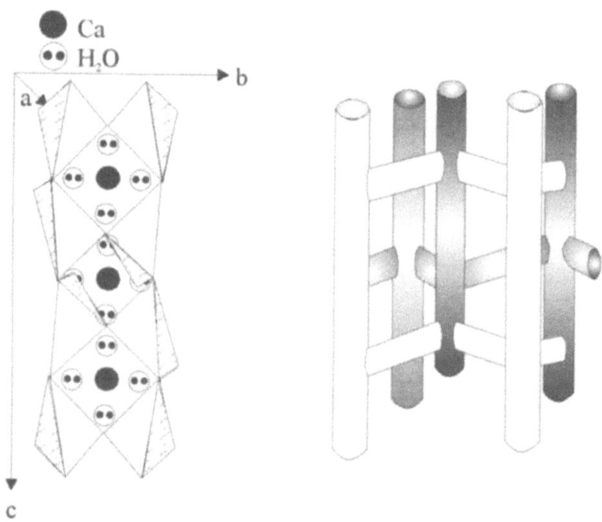

Fig. 5.59: Projection of the Goosecreekite structure [339,340].
Left : View of the framework system in a-direction with cations and water
molecules. Right : The channel system.

Synthesis conditions:

High pressure hydrothermal treatment of synthetic water free glasses of Goosecreekite composition ($2CaO$ x $2Al_2O_3$ x $12SiO_2$).
Hydrothermal parameters: 28 days synthesis time, 1000 bar water pressure, temperature interval 180°C to 250°C.

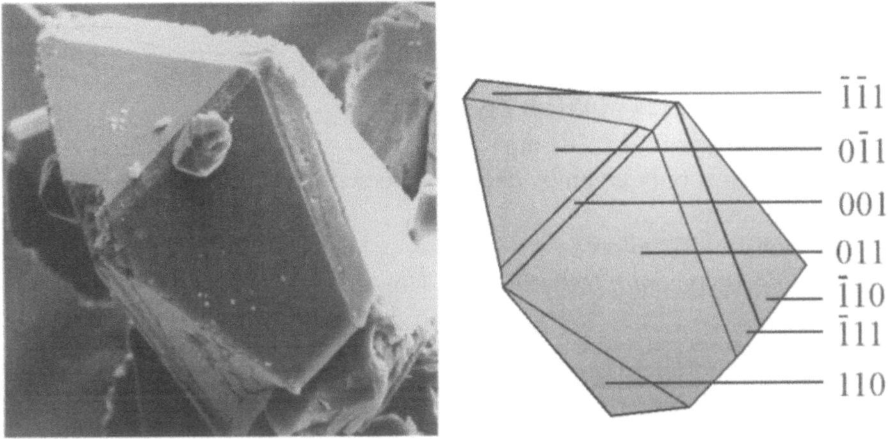

Figure 5.60: Left - the SEM micrographs (357 X) of synthesized Goosecreekite. Right: Indexing of monoclinic Goosecreekite crystal.

Other synthesis approaches: The authors are not aware of other synthesis approaches.

5.5 Zeolites with 5-1 Building Units

5.5.1 Mordenite group

5.5.1.1 Mordenite | $Na^+_2 K^+ Ca^{2+}_2 (H_2O)_{28}$ | $[Al_8Si_{40}O_{96}]$
   – **MOR**: type material

Named after the locality Morden, Nova Scotia, Canada.

Luster: glassy
Channel system(s): [001] **12** 6.5 x 7.0* ↔ {[010] **8** 3.4 x 4.8
                              ↔ [001] **8** 2.6 x 5.7}*
Framework density: 17.2 T/ nm^3
Cages/cavities: only given by channel intersections
Cleavage: none
Color: white, also colored
Crystallographic data: orthorhombic, Cmcm, a = 1.811nm, b = 2.046nm,
                              c = 0.752nm

Hardness: 5
SBU(s): 5-1

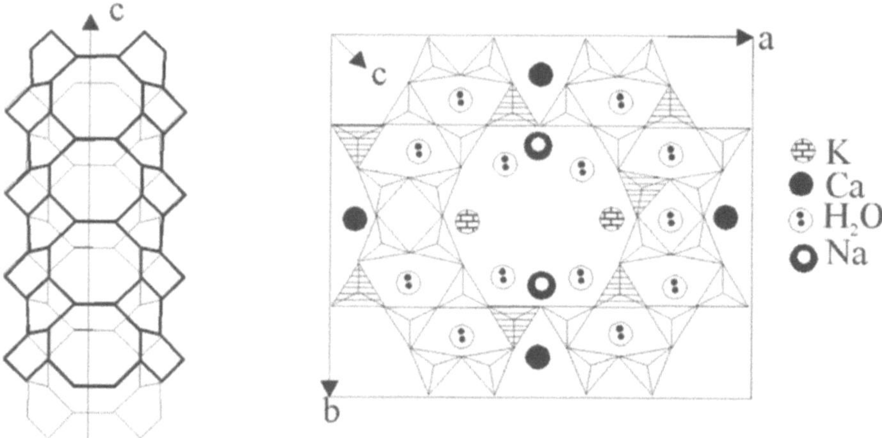

Fig. 5.61: Projection of the Mordenite structure [341,342,343,344,345,346,347,348].
Left : The framework of Mordenite Right : View into the structure from c-
direction with the main channel showing the cations and water molecules.

Synthesis conditions [349]:

High pressure hydrothermal treatment of synthetic water free glasses corresponding the Mordenite composition (MgO, CaO, SrO, BaO, Na_2O, K_2O, Fe_2O_3, Al_2O_3, SiO_2).
Hydrothermal parameters: 60 days synthesis time, 1000 bar water pressure, temperature 220°C.

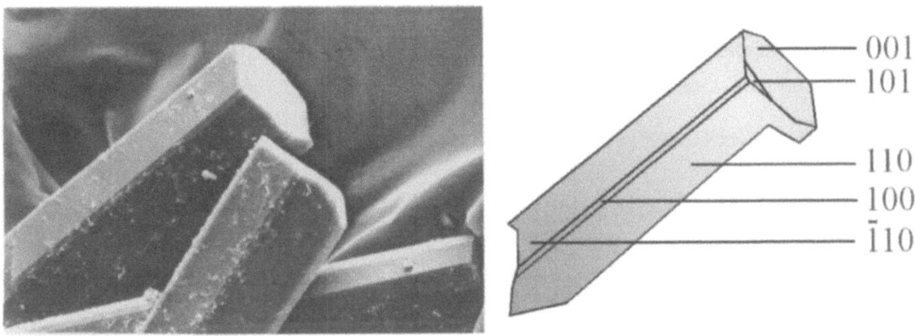

Figure 5.62: Left - the SEM micrographs (500 X) of synthesized Mordenite. Right: Indexing of Mordenite revealing the orthorhombic symmetry.

Other synthesis approaches:

See [102,124,350,351,352,353,354,355,356,357,358,359,360,361,362,363,364,365,366,367,368,369,370, 371,372,373,374,375,376,377,378,379,380,381,382,383,384,385,386,387,388,389,390].

5.5 Zeolites with 5-1 Building Units

5.5.2 Dachiardite group

5.5.2.1 Dachiardite I $(Na^+ K^+ Ca^{2+}_{0.5})_4 (H_2O)_{18}$ I $[Al_4Si_{20}O_{48}]$ – DAC: type material

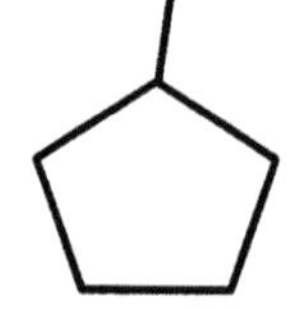

Named after A. D'Achiardi, Italian Mineralogist, University of Pisa

Luster: glassy, translucent
Channel system(s): [010] **10** 3.4 x 5.3* ↔ [001] **8** 3.7 x 4.8*
Framework density: 17.5 T/ nm³
Cages/cavities: only given by channel intersections

Cleavage: according (100) and (001)
Color: colorless
Crystallographic data: monoclinic, C2/m, a = 1.869nm, b = 0.75nm,
c = 1.026nm, β = 107.9°

Hardness: 4 to 4.5
SBU(s): 5-1

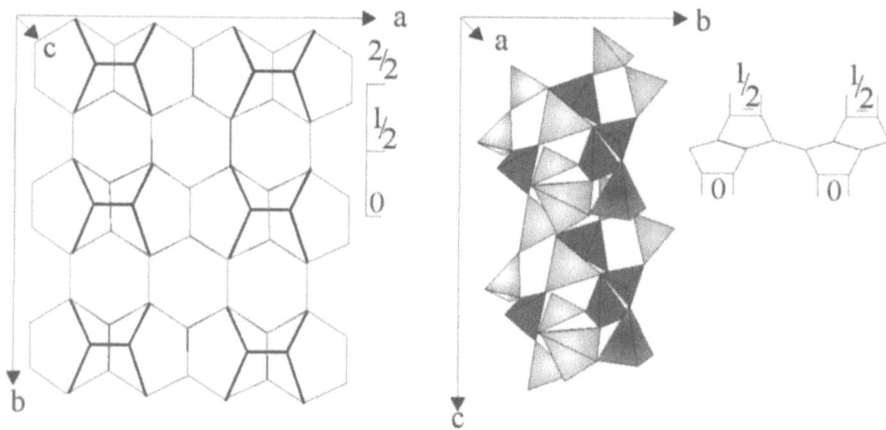

Fig. 5.63: Projection of the Dachiardite structure [391,392,393].
Left : The framework of Dachiardite. Right : The dreier quadruple chain of Dachiardite, right above: The sheet along b.

Synthesis conditions [349]:

High pressure hydrothermal treatment of synthetic water free glasses of Dachiardite composition $(2(Na,K,Ca_{0.5})_2O \times 4Al_2O_3 \times 20SiO_2)$.
Hydrothermal parameters: 60 days synthesis time, 1000 bar water pressure, temperature 220°C.

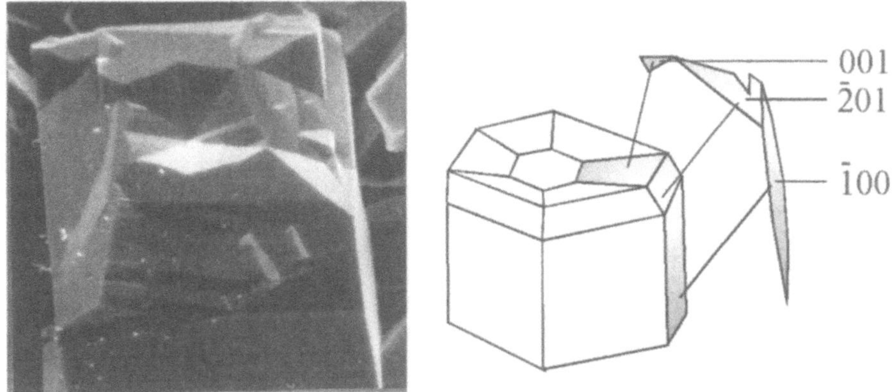

Figure 5.64: Left - the SEM micrograph (1000 X) of synthesized twinned Dachiardite . Right: Indexing of one Dachiardite crystal.

Other synthesis approaches: See [394].

5.5 Zeolites with 5-1 Building Units

5.5.3 Epistilbite group

5.5.3.1 Epistilbite I Ca^{2+}_3 $(H_2O)_{16}$ I $[Al_6Si_{18}O_{48}]$ – EPI:

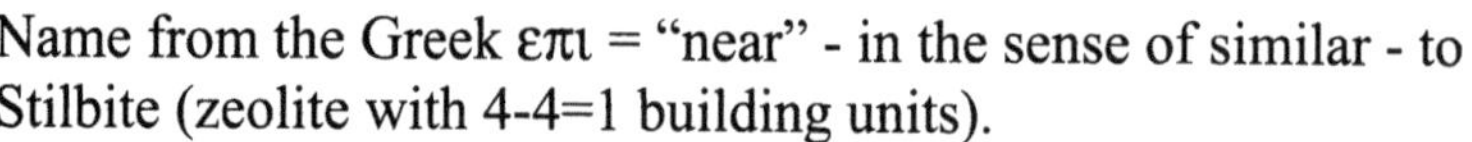

type material

Name from the Greek επι = "near" - in the sense of similar - to
Stilbite (zeolite with 4-4=1 building units).

Remarks: piezoelectric
Luster: glassy
Channel system(s): [100] **10** 3.4 x 5.6* ↔ [001] **8 3.7** x 4.5*
Framework density: 17.6 T/ nm^3
Cages/cavities: only given by channel intersections

Cleavage: according (010)
Color: white, colorless
Crystallographic data: monoclinic, C2/m, a = 0.91nm, b = 1.777nm,
c = 1.022nm, β = 124.6°

Hardness: 4
SBU(s): 5-1

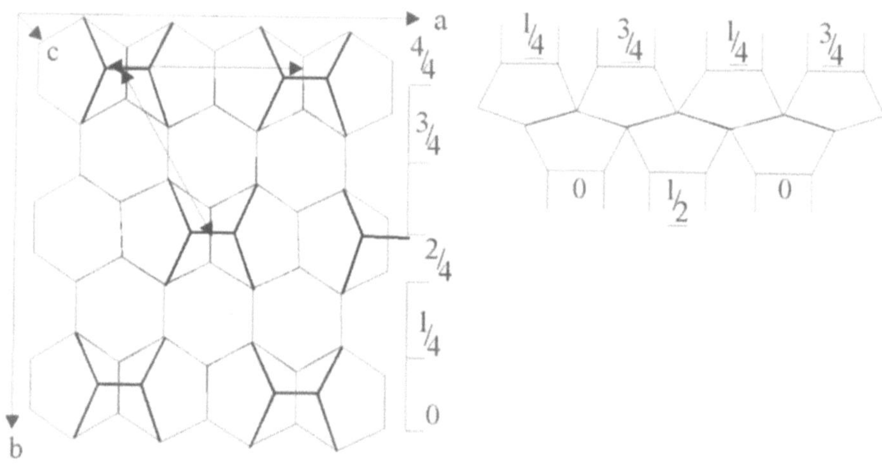

Fig. 5.65: Projection of the Epistilbite structure [395,396,397,398].
Left : The framework of Epistilbite. Right : The sheet arrangement of the epistibite
framework.

Synthesis conditions [349,399]:

High pressure hydrothermal treatment of synthetic water free glasses of Epistilbite composition ($3CaO \times 3Al_2O_3 \times 18SiO_2$).
Hydrothermal parameters: 60 days synthesis time, 1000 bar water pressure, temperature interval 200°C to 230°C.

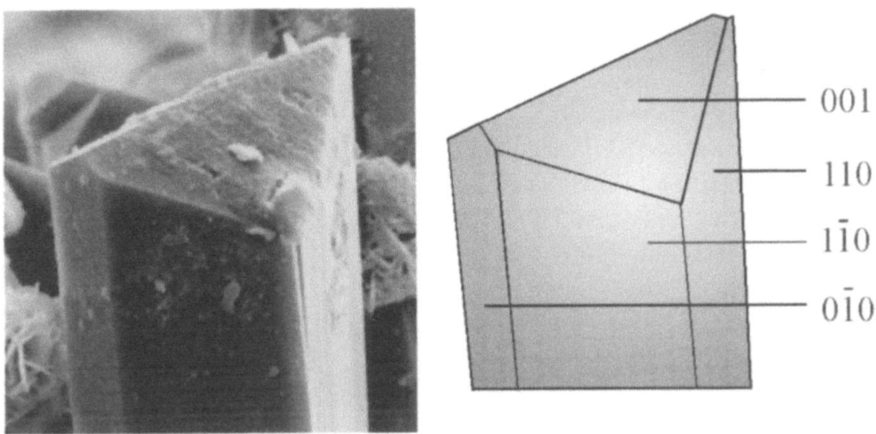

Figure 5.66: Left - the SEM micrographs (1000 X) of synthesized Epistilbite. Right: Indexing of the Epistilbite crystal (monoclinic).

Other synthesis approaches: See [52,53,77,122,133,400,401].

5.5 Zeolites with 5-1 Building Units

5.5.4 Ferrierite group

5.5.4.1 Ferrierite

$| (Na^+, K^+) Mg^{2+}_2 Ca^{2+}_{0.5} (H_2O)_{20} | [Al_6Si_{30}O_{72}] -$ **FER**:

type material

Named after its discoverer W.F. Ferrier, Canadian mineralogist and mining engineer.

Luster: glassy, translucent
Channel system(s): [001] **10** 4.2 x 5.4* ↔ [010] **8** 3.5 x 4.8*
Framework density: 17.8 T/ nm³
Cages/cavities: only given by channel intersections
Cleavage: according (100)
Color: white
Crystallographic data: orthorhombic, Immm, a = 1.918nm, b = 1.414nm,

c = 0.75nm

Hardness: 3.5
SBU(s): 5-1

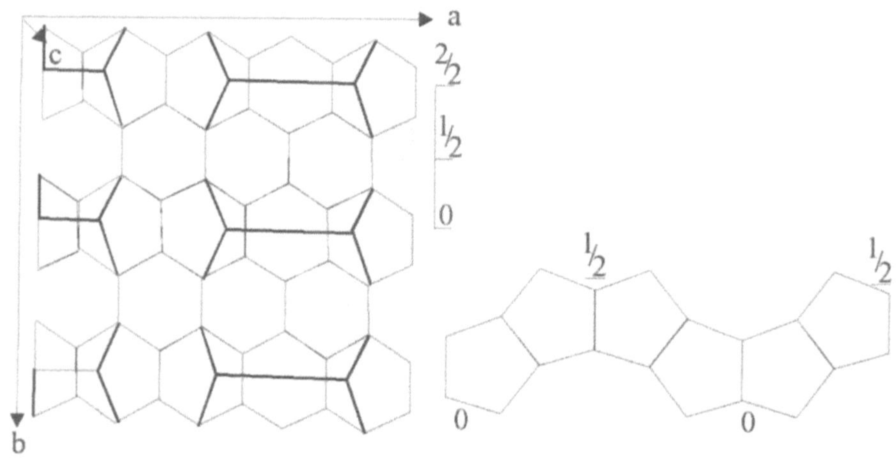

Fig. 5.67: Projection of the Ferrierite structure [402,403,404,405,406,407,408].
Left : The Ferrierite framework. Right : The sheet arrangement of the Ferrierite building units along c.

Synthesis conditions [349]:

High pressure hydrothermal treatment of synthetic water free oxide glasses of Ferrierite composition (1(Na,K)$_2$O x 1CaO x 4MgO x 6Al$_2$O$_3$ x 60SiO$_2$). Hydrothermal parameters: 60 days synthesis time, 1000 bar water pressure, temperature 220°C.

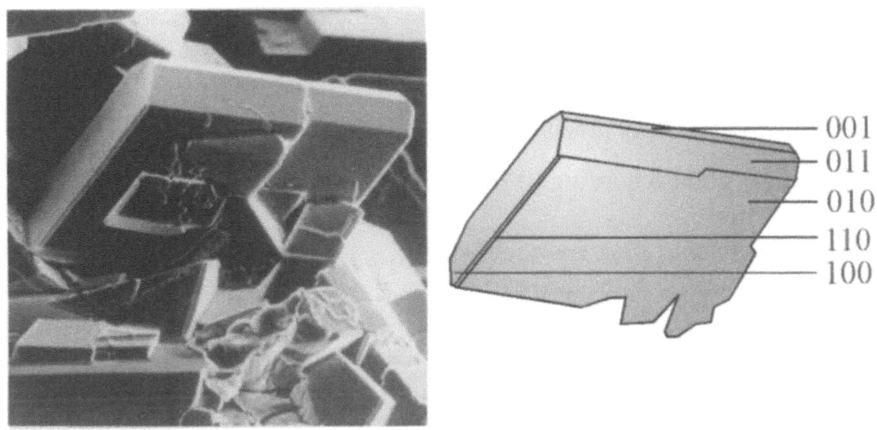

Figure 5.68: Left - the SEM micrograph (1000 X) of synthesized Ferrierite. Right: Indexing of the Ferrierite crystal (orthorhombic).

Other synthesis approaches: See [286,371,404,409,410,384,411,412,413,414,415,416,417,418, 419,420,421,422,423,424,425,426,427,428,429,430,431,432].

5.5 Zeolites with 5-1 Building Units

5.5.5 Bikitaite group

5.5.5.1 Bikitaite I Li^+_2 $(H_2O)_2$ I $[Al_2Si_4O_{12}]$ – **BIK**:
type material

Named after the locality Bikita, Zimbabwe.

Luster: glassy
Channel system(s): [001] **8** 2.8 x 3.7*
Framework density: 20.3 T/ nm^3
Cages/cavities: only given by intersections

Cleavage: good according (001), weak according (100)
Color: white
Crystallographic data: monoclinic, $P2_1$, a = 0.861nm, b = 0.496nm,
c = 0.76nm, β = 114.45°

Hardness: 6
SBU(s): 5-1

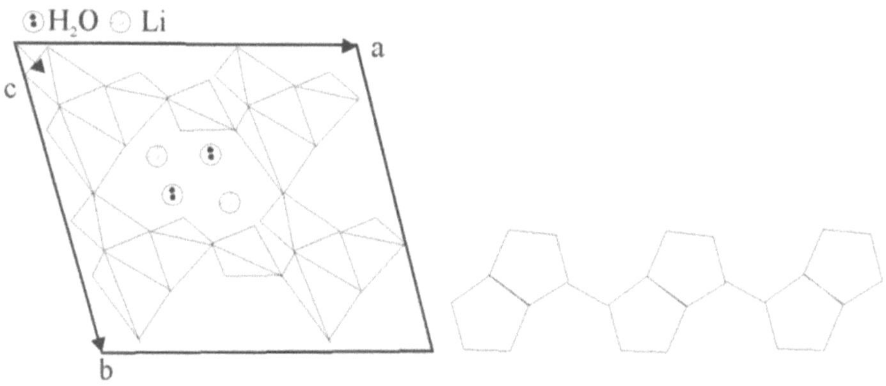

Fig. 5.69: Projection of the Bikitaite structure [433,434,435,436].
Left : Framework with view into the channel system as well as cations and water.
Right : The sheet arrangement of Bikitaite building units.

Synthesis conditions [349]:

High pressure hydrothermal treatment of synthetic water free glasses of Bikitaite composition (Li_2O x $2Al_2O_3$ x $8SiO_2$) in the presence of CO_2. Hydrothermal parameters: 60 days synthesis time, 2000 bar water pressure, temperature 220°C.

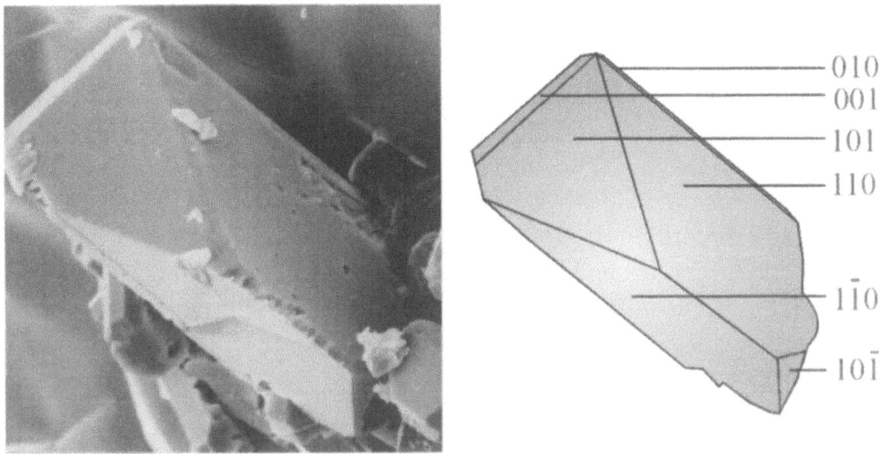

Figure 5.70: Left - the SEM micrographs (422 X) of synthesized Bikitaite. Right: Indexing of the Bikitaite crystal (monoclinic).

Other synthesis approaches: See [437,438].

5.6 Zeolites with 4-4=1 Building Units

5.6.1 Heulandite group

5.6.1.1 Heulandite I (Na$^+$, K$^+$) Ca$^{2+}_4$ (H$_2$O)$_{24}$ I [Al$_9$Si$_{27}$O$_{72}$]
– HEU: type material

Named after H. Heuland, English mineral collector.

Luster: glassy, pearly on (010)
Channel system(s): {[001] **10** 3.1 x 7.5* + **8** 3.6 x 4.6*}
　　　　　　　　　　↔ [100] **8** 2.8 x 4.7*
　　　　　　　　　　Variable in size (considerable flexibility of framework)
Framework density: 17.1T/ nm^3
Cages/cavities: only given by channel intersections
Cleavage: according (010)
Color: colorless, white, brick-red
Crystallographic data: monoclinic, C2/m, a = 1.770nm, b = 1.794nm,
　　　　　　　　　　　　c = 0.742nm, β = 116.4°

Hardness: 3.5 to 4
SBU(s): 4-4=1

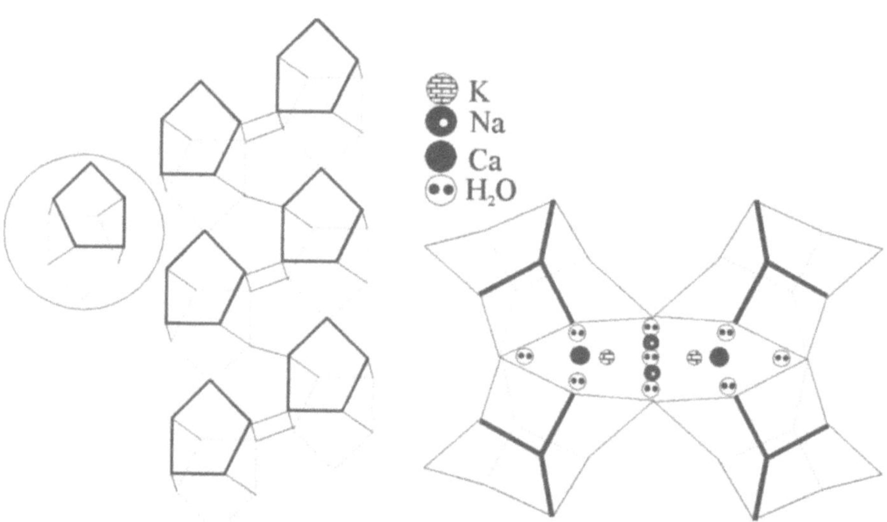

Fig. 5.71: Projection of the Heulandite structure[439,440,441,442,443,444,445,446,447,448,449,450,451].

Left: View along b. Right: The channel system in c-direction with cations and water molecules.

Synthesis conditions [452]:

High pressure hydrothermal treatment of synthetic water free glasses of Heulandite composition ($(Na,K)_2O$ x $1MgO$ x $1SrO$ x $1BaO$ x $1CaO$ x $4.5Al_2O_3$ x $27SiO_2$) containing 5mol% of Fe_2O_3 as a kind of structure directing catalyst.

Hydrothermal parameters: 60 days synthesis time, 1000 bar water pressure, temperature interval 120°C to 400°C.

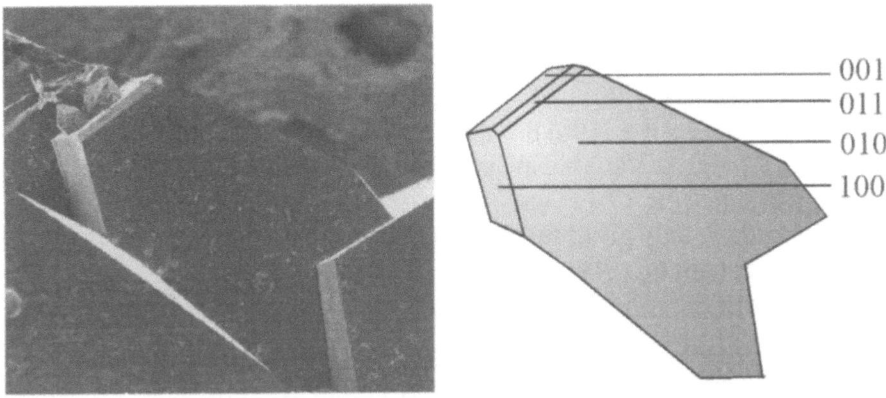

Figure 5.72: Left - the SEM micrographs (90 X) of synthesized Heulandite with second crystal generation on $(0\bar{1}0)$. Right: Indexing of monoclinic Heulandite crystal.

Other synthesis approaches: See [52,53,122,123,133,209,211,453,454,455,456,457].

5.6 Zeolites with 4-4=1 Building Units

5.6.1 Heulandite group

5.6.1.2 Clinoptilolite I $(Na^+, K^+)_6$ $(H_2O)_{20}$ I $[Al_6Si_{30}O_{72}]$
–   HEU:

Name from the Greek κλινοσ = inclined and πτιλοσ = dusk, in the sense of "monoclinic Ptilolite (Mordenite)"

Remarks: Heulandite and Clinoptilolite are different in their Si/Al ratio: Si/Al < 4 in Heulandite.

Luster: glassy, translucent
Channel system(s): {[001] **10** 3.1 x 7.5* + **8** 3.6 x 4.6*}
                              ↔ [100] **8** 2.8 x 4.7*
                              Variable in size (considerable flexibility of framework)
Framework density: 17.1T/ nm^3
Cages/cavities: only given by channel intersections
Cleavage: according (010)
Color: colorless

Crystallographic data: monoclinic, C2/m, a = 1.762nm, b = 1.791nm,
                              c = 0.739nm, β = 116.266°

Hardness: 3.5
SBU(s): 4-4=1

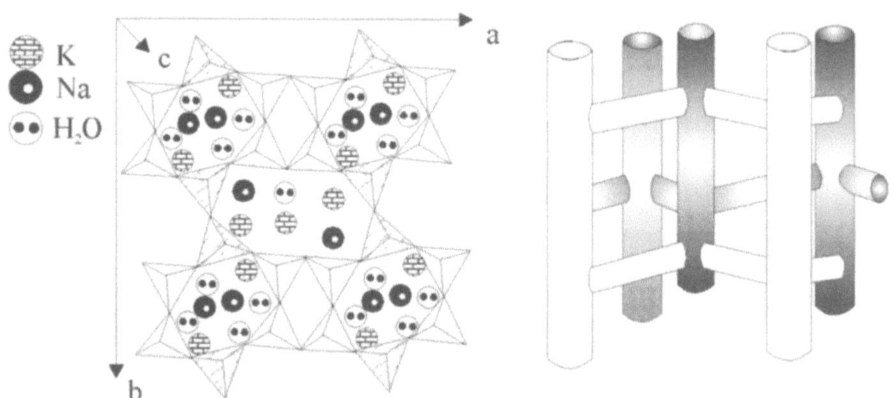

Fig. 5.73: Projection of the Clinoptilolite structure [458,459,460,461,462,463,464].
Left: View along b. Right: The channel system in c-direction with cations and water molecules.

Synthesis conditions:

High pressure hydrothermal treatment of synthetic water free glasses of Clinoptilolite composition (5(Na,K)$_2$O x 0.25MgO x 0.25SrO x 0.25BaO x 0.25CaO x 3Al$_2$O$_3$ x 30SiO$_2$) containing 5mol% of Fe$_2$O$_3$ as a kind of structure directing catalyst.
Hydrothermal parameters: 60 days synthesis time, 1000 bar water pressure, temperature interval 120°C to 400°C.

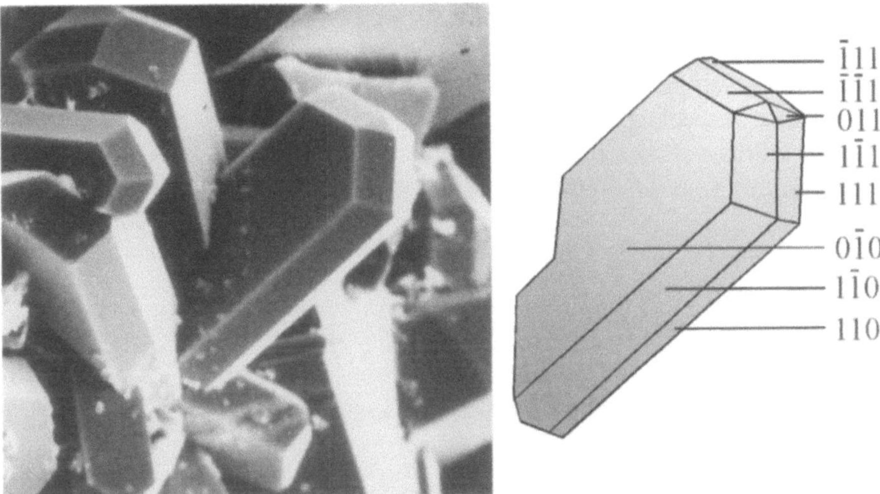

Figure 5.74: Left - the SEM micrograph (1000 X) of synthesized Clinoptilolite. Right: Indexing of the monoclinic Clinoptilolite crystal.

Other synthesis approaches: See [52,124,453,465,466,467,468,469,470,471,472,473,474].

5.6 Zeolites with 4-4=1 Building Units

5.6.2 Stilbite group

5.6.2.1 Stilbite I Na$^+$ Ca$^{2+}_4$ (H$_2$O)$_{30}$ I [Al$_9$Si$_{27}$O$_{72}$]
— **STI**: type material

Name from the Greek στιλβπ = luster

Luster: glassy
Channel system(s): [100] **10** 4.7 x 5.0* ↔ [001] **8** 2.7 x 5.6*
Framework density: 16.3 T/ nm^3
Cages/cavities: only given by channel intersections
Cleavage: according (010)
Color: white, yellowish, reddish
Crystallographic data: monoclinic, C2/m, a = 1.361nm, b = 1.824nm,
c = 1.127nm, β = 127.85°

Hardness: 3.5 to 4
SBU(s): 4-4=1

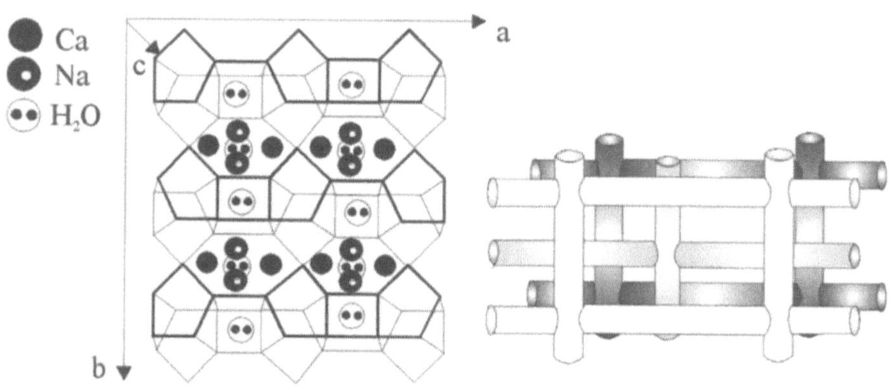

Fig. 5.75: Projection of the Stilbite structure [475,476,477,478,479,480,481].
View in c-direction showing the framework as well as the channel system with cations and water molecules.

Synthesis conditions [482,483]:

High pressure hydrothermal treatment of synthetic water free glasses of Stilbite composition containing up to 5mol% Fe_2O_3, MgO and K_2O.
Hydrothermal parameters: up to 60 days synthesis time, 1000 bar water pressure, temperature interval of synthesis: 155°C to 400°C.

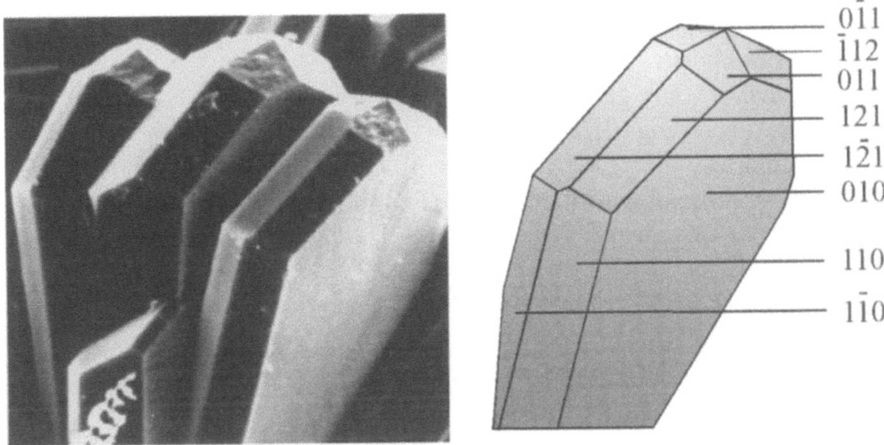

Figure 5.76: Left - the SEM micrographs (2000 X) of synthesized Stilbite. Right: Indexing of Stilbite crystal (monoclinic).

Other synthesis approaches: See [484,485].

5.6 Zeolites with 4-4=1 Building Units

5.6.2 Stilbite group

5.6.2.2 Stellerite I Ca^{2+}_4 $(H_2O)_{16}$ I $[Al_8Si_{28}O_{72}]$ – STI

Named after W. Steller, naturalist from Iceland.

Luster: translucent
Channel system(s): [100] **10** 4.7 x 5.0* ↔ [001] **8** 2.7 x 5.6*
Framework density: 16.3 T/ nm^3
Cages/cavities: only given by channel intersections
Cleavage: according (010)
Color: colorless
Crystallographic data: orthorhombic, Fmmm, a = 1.36nm, b = 1.822nm,
$\qquad\qquad\qquad\qquad\qquad\qquad\qquad\qquad$ c = 1.784nm

Hardness: 4
SBU(s): 4-4=1

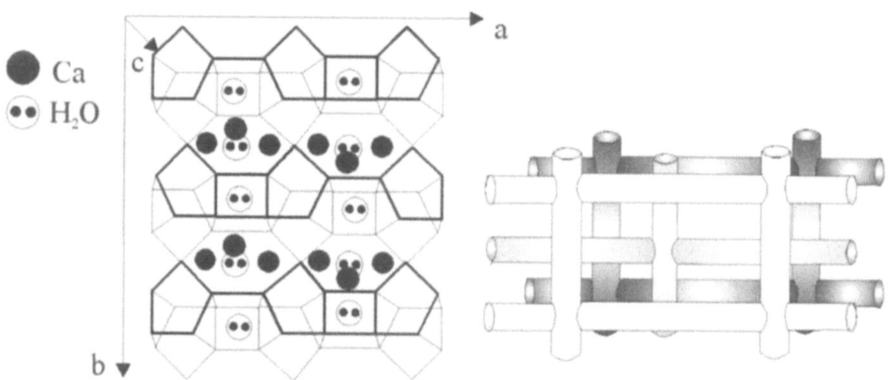

Fig. 5.77: Projection of the Stellerite structure [486,487,488,489,490,491] in c-direction showing the framework and the cations.

Synthesis conditions [483]:

High pressure hydrothermal treatment of synthetic water free glasses of Stellerite composition containing up to 5mol% Fe_2O_3, MgO and K_2O. Hydrothermal parameters: up to 60 days synthesis time, 1000 bar water pressure, temperature interval of synthesis: 200°C to 250°C.

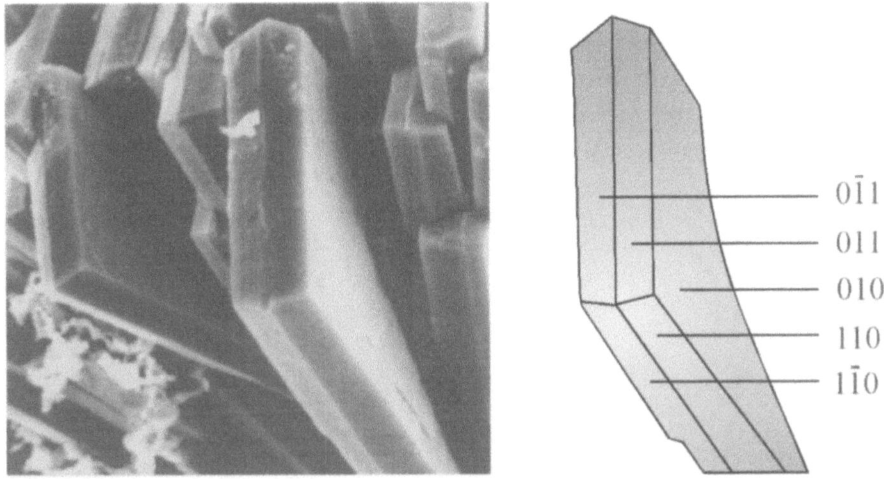

Figure 5.78: Left - the SEM micrographs (2000 X) of synthesized Stellerite. Right: Indexing of Stellerite crystal (orthorhombic).

The authors are not aware of other synthesis approaches.

5.6 Zeolites with 4-4=1 Building Units

5.6.2 Stilbite group

5.6.2.3 Barrerite I Na$^+_8$ (H$_2$O)$_{26}$ I [Al$_8$Si$_{28}$O$_{72}$] – STI

Named after R.M. Barrer, famous zeolite specialist.

Luster: translucent
 Channel system(s): [100] **10** 4.7 x 5.0* ↔ [001] **8** 2.7 x 5.6*
Framework density: 16.3 T/ nm^3
Cages/cavities: only given by channel intersections
Cleavage: according (010)
Color: white
Crystallographic data: orthorhombic, Fmmm, a = 1.364nm, b = 1.82nm,
                          c = 1.784nm
Hardness: 4
SBU(s): 4-4=1

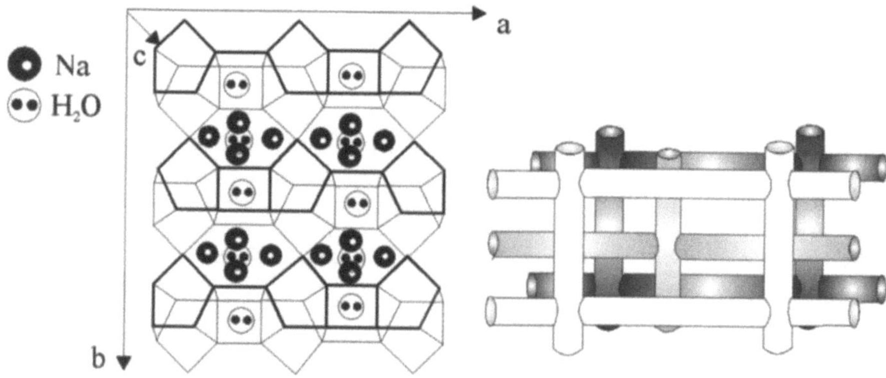

Fig. 5.79: Projection of the Barrerite structure [492,493,494,495].
The framework is identical to that of Stellerite, the difference lies in the cations.

Synthesis conditions [483]:

High pressure hydrothermal treatment of synthetic water free glasses of Barrerite composition containing up to 5mol% Fe_2O_3, MgO and K_2O. Hydrothermal parameters: up to 60 days synthesis time, 1000 bar water pressure, temperature interval of synthesis: 200°C to 250°C.

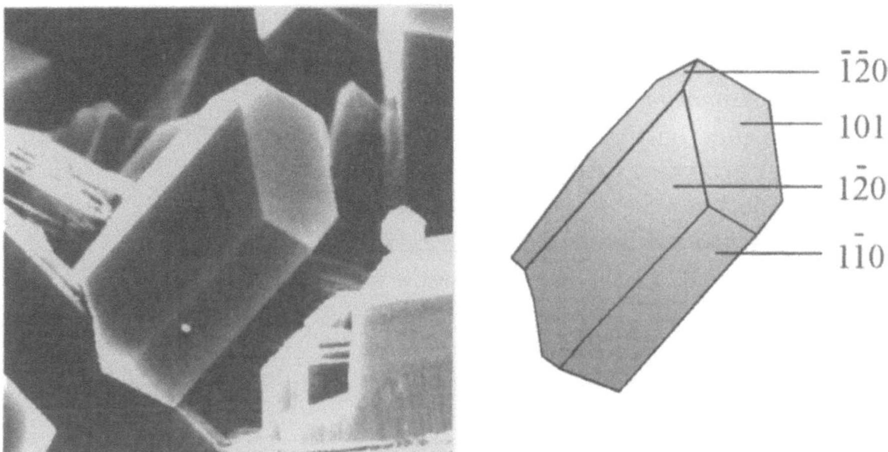

Figure 5.80: Left - the SEM micrographs (2000 X) of synthesized Barrerite. Right: Indexing of Barrerite crystal (orthorhombic).

The authors are not aware of other synthesis approaches.

5.6 Zeolites with 4-4=1 Building Units

5.6.3 Brewsterite group

5.6.3.1 Brewsterite I Sr^{2+}_2 $(H_2O)_{10}$ I $[Al_4Si_{12}O_{32}]$
   BRE: type material

Named after D. Brewster, Scottish naturalist.

Luster: glassy
Channel system(s): [100] **8** 2.3 x 5.0* ↔ [001] **8** 2.8 x 4.1*
Framework density: 17.3 T/ nm^3
Cages/cavities: only given by channel intersections
Cleavage: according (010)
Color: colorless
Crystallographic data: monoclinic, P2$_1$/m, a = 0.677nm,
                         b = 1.751nm, c = 0.774nm, β = 94.3°

Hardness: 3
SBU(s): 4-4=1

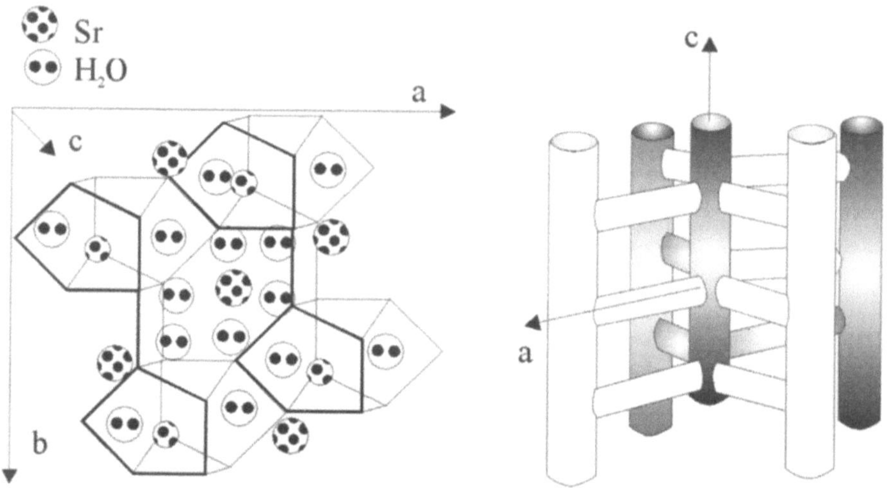

Fig. 5.81: Projection of the Brewsterite structure [496,497,498,499,500,501].
Left : View of the framework from c-direction with cations and water molecules.

Synthesis conditions [502]:

High pressure hydrothermal treatment of synthetic water free glasses of Brewsterite composition ($2SrO \times 2Al_2O_3 \times 12SiO_2$).
Hydrothermal parameters: 60 days synthesis time, 2000 bar water pressure, temperature 180° to 250°C.

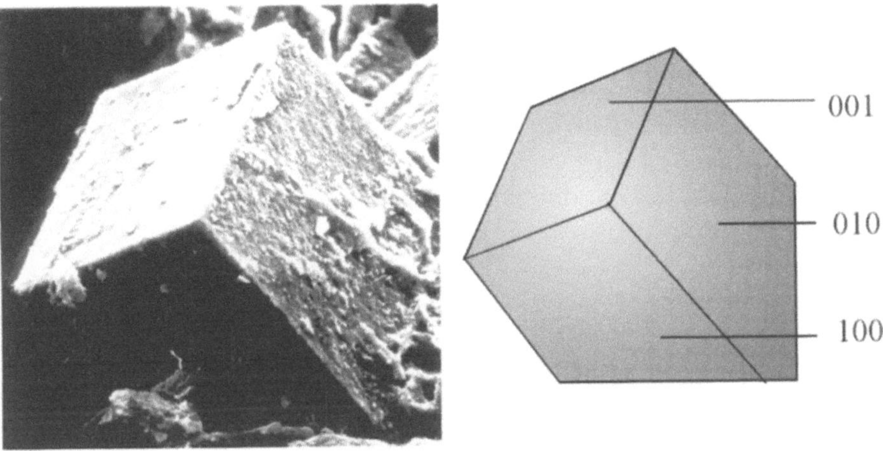

Figure 5.82: Left – the SEM micrographs (1000 X) of synthesized Brewsterite. Right: Indexing of Brewsterite crystal (orthorhombic).

Other synthesis approaches: See [122,165,400,503].

5.7 Zeolites with Unknown Structure Types

5.7.1 Cowlesite group

5.7.1.1 Cowlesite I Ca^{2+}_6 $(H_2O)_{36}$ I $[Al_{12}Si_{18}O_{60}]$ – COW

Named after J. Cowles, mineral collector, Oregon, USA.

Luster: translucent
Channel system(s): unknown
Framework density: unknown
Cages/cavities: unknown
Cleavage: according (010)
Color: colorless
Crystallographic data: orthorhombic, a = 1.127nm, b = 1.525nm,
                                                c = 1.261nm

Hardness: 4
SBU(s): unknown

Synthesis conditions :

High pressure hydrothermal treatment of synthetic water free glasses of Cowlesite composition (2CaO,MgO, Na_2O, K_2O, $2Al_2O_3$ x $12SiO_2$); Fe_2O_3 serves as a sort of structure directing agent.
Hydrothermal parameters: 28days synthesis time, 1000 bar water pressure, temperature 180°C to 250°C.

Figure 5.84: Left – the SEM micrographs (2000 X) of synthesized Cowlesite. Right: Indexing of the Cowlesite crystals (orthorhombic!).

Other synthesis approaches: See [504].

6. CONCLUDING REMARKS

From all natural zeolites introduced by Gottardi and Galli, Parthéite is the only one we were not able to synthesize. Under all circumstances it can not be synthesized by using precursor glasses of the proper zeolite composition.
The experience we have gained during the zeolite construction work makes us confident that this problem is only temporary and will be solved by adding further oxides systematically as doping materials.

Much more challanging tasks, however, are the reconstruction of zeolites which have been synthesized by other hydrothermal techniques before they have been found in nature. The comparison of their physical and chemical properties (i.e. Bronstedt site activity, structural stability) will have to reveal the differences between materials obtained by the low- and high pressure hydrothermal synthesis access and that these differences are small details but systematic ones.

The hydrothermal method under pressure has been shown to be an excellent synthesis tool in order to implement transition metal cations into the zeolite framework.
We are convinced that the research for new microporous materials should use this tool in a systematic way.

7. ACKNOWLEDGMENT

The authors gratefully acknowledge the support from the French Centre National de la Recherche Scientifique, CNRS.

8. INDEX

9. REFERENCES

[1] G. Gottardi, E. Galli : Natural Zeolites, Springer, Berlin, Heidelberg, New York, Tokyo 1985.

[2] R.M. Barrer : The Hydrothermal Chemistry of Zeolites, Academic Press, London 1982.

[3] A.F. Cronstedt, Kongl. Vetenskaps, Acad. Handl. Stockh. 17, 120 (1756).

[4] L.B. McCusker, F. Liebau and G. Engelhardt, Pure Appl. Chem. 73,(2) 381-394, (2001).

[5] Ch. Baerlocher, W.M. Meier, D.H. Olson, Atlas of Zeolite Framework Types, 5th rev. edn., Elsevier Amsterdam, London, New York, Oxford, Paris, Shannon, Tokyo (2001).

[6] D.S. Coombs, A. Alberti, T. Armbruster, G. Artioli, C. Colella, E. Galli, J.D. Grice, F. Liebau, J.A. Mandarino, H. Minato, E.H. Nickel, E. Passaglia, D. Peacor, S. Quartieri, R. Rinaldi, M. Ross, R.A. Sheppard, E. Tillmanns, G. Vezzalini, The Canadian Mineralogist, 35, 1571-1606, (1997).

[7] O. Schäf, H. Ghobarkar, Recent Research Reports 01-R-01 13th International Zeolite Conference Montpellier July 8-13, Groupe Français des Zéolithes (2001).

[8] H. Ghobarkar, Cryst. Res. Technol. 27, 1071-1075 (1992).

[9] P.A. Jacobs, J.A. Martens, Synthesis of High-Silica Aluminosilicate Zeolites. Studies in Surface Science and Catalysis Vol. 33, Elsevier, Amsterdam, Oxford, New York, Tokyo (1987).

[10] J. Weitkamp, L. Puppe, Catalysis and Zeolites, Fundamentals and Application, Springer Verlag, Berlin, Heidelberg, New York, London, Paris, Tokyo (1999).

[11] H.G. Karge, J. Weitkamp, Molecular Sieves, Science and Technology, Vol.1, Springer Verlag, Berlin, Heidelberg, New York, London, Paris, Tokyo (1998).

[12] H. Robson, Verified Syntheses of Zeolitic Materials, 2nd rev. edn., Elsevier Amsterdam, London, New York, Oxford, Paris, Shannon, Tokyo (2001).

[13] FIZ Karlsruhe and NIST Gaithersburg, Inorganic Crystal Structure Database ICSD, release 2 (2002).

[14] L.B. McCusker, F. Liebau, G. Engelhardt, Microporous and Mesoporous Materials 58, 3-13, (2003).

[15] E. Galli, Rend. Soc. Ital. Miner. Petrol. 31, 549-564, (1975).

[16] W.J. Mortimer, Compilation of Extra Framework Sites in Zeolites, IZA publication, Elsevier, London, Boston, Singapore, Sydney, Toronto, Wellington (1982).

[17] O. Schäf, H. Ghobarkar, P. Knauth, Hydrothermal Synthesis of Nanomaterials, in Nanostructured Materials : Selected Synthesis Methods, Properties and Applications, P. Knauth and J. Schoonman ed., Kluwer, Boston p. 24, (2003).

[18] K. Byrappa, M. Yoshimura, I. Yoshimura (eds.), Handbook of Hydrothermal Technology: Technology for Crystal Growth and Materials Processing, William Andrew Publishing (2001).

[19] J. Jung, M. Perrut, J. Supercrit. Fluids 20, 179-219, (2001).

[20] J.-P. Jolivet, Condensation et précipitation en phase aqueuse, in De la solution à l'oxyde, CNRS éditions, Paris p. 34-67, (1994).

[21] G. Kortüm, Treatise on Electrochemistry, Elsevier, Amsterdam p. 105, (1965).

[22] A. R. West, Basic Solid State Chemistry, J. Wiley, New York p. 75, (1994).

[23] P.W. Atkins, Physical Chemistry 6th edn., Freeman and Co. (1997).

[24] M.M. Hoffmann, JG. Darab, B.J. Palmer, J.L. Fulton, J. Phys. Chem. A 42, 8471-4882, (1999).

[25] J.L. Fulton, D.M. Pfund, S.L. Wallen, M. Newville, E.A. Stern, Y. Ma, J. Chem. Phys. 105, 2161-2166, (1996).

[26] H. Ghobarkar, O. Schäf and U. Guth, Prog. Solid St. Chem. 27(2-4), 29-73 (1999).

[27] O. A. Bogatikov (ed.), R. F. Fursenko (transl.) Magmatism and Geodynamics: Terrestrial Magmatism Throughout the Earth's History, Taylor and Francis, (2000).

[28] H.G.F. Winkler, Petrogenesis of Metamorphic Rocks, 5th edn., Springer, New York, Heidelberg, Berlin (1979).

[29] H. Ghobarkar, Ph.D. Thesis, Free University of Berlin, (1978).

[30] H. Ghobarkar, Krist. Tech. 12, K49-51, (1977) .

[31] D.S. Coombs, A. Alberti, T. Armbruster, G. Artioli, C. Colella, E. Galli, J.D. Grice, F. Liebau, J.A. Mandarino, H. Minato, E.H. Nickel, E. Passaglia, D. Peacor, S. Quartieri, R. Rinaldi, M. Ross, R.A. Sheppard, E. Tillmanns, G. Vezzalini, The Canadian Mineralogist, 35, 1571-1606, (1997).

[32] G.Y. Chao, Can. Mineral. 18, 85-88, (1980).

[33] T.T. Chen, G.Y. Chao, Can. Mineral. 18, 77-84, (1980).

[34] A. Kirfel, M. Orthen, G. Will, Zeolites 4, 140-146 (1984).

[35] W.H. Baur, W. Joswig, Neues Jahrbuch fuer Mineralogie. Monatshefte 171-187 (1996).

[36] E. Meneghinello, A. Martucci, A. Alberti, F. Di Renzo, Microporous and Mesoporous Materials 30, 89-94 (1999).

[37] H.T. Evans jr., J.A. Konnert, M. Ross, American Mineralogist 85, 1808-1815 (2000).

[38] H. Ghobarkar, O. Schäf, Cryst. Res. Technol. 31, K54-57 (1996).

[39] E.E. Senderov, N.I. Khitarov, Molecular Sieve Zeolites-I, Advan. Chem. Ser., 101, 149-154 (1971).

[40] E.E. Senderov, N.I. Khitarov, Geokhimiya 12, 1398-412, (1966).

[41] H. Höller, U. Wirsching in : LVC. Rees (ed., Proc. 5th Int. Conf. Zeol., London 164-170, (1980).

[42] G. Artioli, J.V. Smith, J.J. Pluth, Acta Crystallographica C 42, 937-942, (1986).

[43] K. Stahl, R. Thomasson, Zeolites 14, 12-17 (1994).

[44] E. Stuckenschmidt, A. Kirfel, European Journal of Mineralogy 12, 571-579 (2000).

[45] H. Ghobarkar, O. Schäf, Cryst. Res. Technol. 31(6), K67-K69 (1996).

[46] L. Faelth, S. Hansen, Acta Crystallographica B 35, 1877-1880, (1979).

[47] W. Joswig, H. Bartl, H. Fuess, Zeitschrift fuer Kristallographie 166, 219-223, (1984).

[48] A. Kvick, K. Stahl, J.V. Smith, Zeitschrift fuer Kristallographie 171, 141-154, (1985).

[49] E. Stuckenschmidt, W. Joswig, W.H. Baur, W. Hofmeister, Physics and Chemistry of Minerals 24, 403-410, (1997).

[50] S. Kuntzinger, N.E. Ghermani,Y. Dusausoy, C. Lecomte, Acta Crystallographica B 54, 819-833, (1998).

[51] H. Ghobarkar, O. Schäf, Cryst. Res. Technol. 31(6), K70-K72 (1996).

[52] U. Barth-Wirsching, H. Hoeller, Eur. J. Mineral. 1(4), 489-506 (1989).

[53] M. Koizumi, R. Roy, J. Geol. 68, 41-53 (1960).

[54] F. Mazzi, A.O. Larsen, G. Gottardi, E. Galli, Neues Jahrbuch fuer Mineralogie. Monatshefte 1986, 219-228 (1986).

[55] G. Artioli, M.R. Torres Salvador, Materials Science Forum 79, 845-850 (1991).

[56] G. Artioli, E. Galli, American Mineralogist 84, 1445-1450 (1999).

[57] H. Ghobarkar, O. Schäf, Zeolites 19(4), 259-261, (1997).

[58] M. Aoki, J. Japan. Assoc. Miner. Petr. Econ. Geol. 73,155-166, (1978).

[59] E. Galli, Acta Crystallographica B 32, 1623-1627, (1976).

[60] I.A. Belitsky, S.P. Gabuda, W. Joswig, H. Fuess, Neues Jahrbuch fuer Mineralogie. Monatshefte 1986, 541-551, (1986).

[61] F. Mazzi, E. Galli, G. Gottardi, Neues Jahrbuch fuer Mineralogie. Monatshefte 1984, 373-382 (1984).

[62] A. Kvick, J.V. Smith, Journal of Chemical Physics 79, 2356-2362 (1983).

[63] H. Ghobarkar, O. Schäf, Cryst. Res. Technol. 32(5), 653-657, (1997).

[64] R.M. Barrer, J.W. Baynham, J. Chem. Soc. London, 2882-2891, (1956).

[65] C. Colella, M. De'Gennaro, V. Iorio, Stud. Surf. Sci. Catal. 28 263-70, (1986).

[66] C. Colella, Rend. Accad. Sci. Fis. Mat., Naples 40, 303-13 (1973).

[67] J.D. Sherman, ACS Sym. Ser. 40, 30-42, (1977).

[68] H.J. Bosmans, E. Tambuyzer, J. Paenhuis, L. Ylen, J. Vancluisen in : W.M. Meier, J.B. Ujetterhoeven (eds.) Molecular Sieves, Am. Chem. Soc. Adv. Chem. Ser. 121, 179-188, (1973).

[69] H.E. Robson and K.P. Lillerud (eds.), Verified Syntheses of Zeolitic Materials, 2nd rev. Edn., Elsevier, Amsterdam, London, N.Y. Oxford, Paris, Shannon, Tokyo 2001.

[70] S.T. Amirov, I.R. Amiraslanov, B.T. Usubaliev, H.S. Mamedov, Azerbaidzhanskii Khimicheskii Zhurnal 1978, 120-127, (1978).

[71] W.H. Taylor, C.A. Meek, W.W. Jackson, Zeitschrift fuer Kristallographie, Kristallgeometrie, Kristallphysik, Kristallchemie 84, 373-398, (1933).

[72] F. Pechar, Crystal Research and Technology 17, 1141-1144, (1982).

[73] J.J. Pluth, S.V. Smith, A. Kvick, Zeolites 5, 74-80, (1985).

[74] K. Stahl, A. Kvick, J.V. Smith, Acta Crystallographica C 46, 1370-1373, (1990).

[75] A. Alberti, G. Vezzalini, V. Tazzoli, Zeolites 1, 91-97, (1981).

[76] A. Yamazaki, Y. Inoue, M. Koike, T. Sakamoto, R.J. Otsuka, Therm. Anal. 40(1), 85-97, (1993).

[77] R.M. Barrer, P.J. Denny, J. Chem. Soc. 983-1000, (1961).

[78] E.G. Ehlers, J. Geol. 61, 231-51, (1953).

[79] J.R. Goldsmith, Mineralog. Mag. 29, 952-959, (1952).

[80] V.C. Juan, H.J. Lo, Proc. Geol. Soc. (Taiwan), 12, 21-29 (1969).

[81] U. Wirsching, Clays and Clay Minerals, 29, 171-183, (1983).

[82] H. Ghobarkar, O. Schäf, Materials Science and Engineering B60, 163-167, (1999)

[83] F. Mazzi, E. Galli, American Mineralogist, 63, 448-460, (1978).

[84] G. Ferraris, D.W. Jones, J. Yerkess, Zeitschrift fuer Kristallographie, Kristallgeometrie, Kristallphysik, Kristallchemie 135, 240-252, (1972).

[85] H. Ghobarkar, W. Franke, Cryst. Res. Technol. 21(8) 1071-1075, (1986).

[86] G. Kirov, V. Pechigargov, E. Landzheva, Chem. Geol. 26(1-2), 17-28, (1979).

[87] M. Koizumi, Hwahak Kwa Kongop Ui Chinbo, 19(2), 81-5, (1979).

[88] S. Ueda, M. Koizumi, Am. Mineral. 64(1-2), 172-9, (1979).

[89] H. Abe, M. Aoki, Chem. Geol. 17(2), 89-100, (1976).

[90] E.E. Senderov, Bull. Soc. Fr. Mineral. Cristallogr. 97(2-5), 393-402, (1974).

[91] J.R. Boles, Amer. Mineral. 56(9-10), 1724-1734, (1971).

[92] K.T. Kim , B.J. Burley, Miner. Mag. 43, 1035-1045, (1980).

[93] M.L. Sand, W.S. Coblenz, L.B. Sand, Advan. Chem. Ser. 101 (Molecular Sieve Zeolites-I), 127-34, (1971).

[94] F. Pechar, Acta Univ. Carol., Geol. 2, 165-179, (1990).

[95] H.J. Lo, Chung-kuo Ti Chih Hsueh Hui Chuan K'an 8, 87-94, (1987).

[96] M.S. Joshi, A.L. Choudhari, R. Kanitkar, Cryst. Res. Technol. 18(11), 1347-1351, (1983).

[97] H. Ghobarkar, O. Schäf, P. Knauth, Ann. Chim. Sci. Mat., 24, 209-215, (1999).

[98] Y. Takeuchi, F. Mazzi, N. Haga, E. Galli, American Mineralogist 64, 993-1001, (1979).

[99] C.M.B. Henderson, A.M.T. Bell, S.C. Kohn, C.S. Page, Mineralogical Magazine 62, 165-178, (1998).

[100] H. Ghobarkar, Cryst. Res. Technol. 20(9), K90-K92, (1985).

[101] D.S. Coombs, A.J. Ellis, W.S. Fyte, A.M. Taylor, 17, 53-107, (1959).

[102] L.L. Ames, L.B. Sand, Am. Mineralogist 43, 476-80, (1958).

[103] R.K. Rastsvetaeva, O.Yu. Rekhlova, V.I. Andrianov, Yu.A. Malinovskii, Doklady Akademii Nauk SSSR, 316, 624-628, (1991).

[104] S. Ueda, M. Koizumi, Advan. Chem. Ser. 101 (Molecular Sieve Zeolites-I), 135-139, (1971).

[105] F. DiRenzo, Z. Gabelica in : Natural Zeolites '93 : Occurrence, Properties, Use. D.W. Ming, F.A. Mumpton (eds.). Int. Committee on Natural Zeolites, Brockport, New York, 173-185, (1995).

[106] Y. Kim, R.J. Kirkpatrick, Mineralogical Magazine 60(403), 957-962 (1996).

[107] W.D. Balgord, R. Roy, Advan. Chem. Ser. 101(Molecular Sieve Zeolites-I), 140-148, (1971).

[108] R.M. Barrer, D.J. Marshall, J. Chem. Soc. 11, 6616-6621, (1965).

[109] T. Armbruster, T. Kohler, Neues Jahrbuch fuer Mineralogie. Monatshefte 1992, 385-397, (1992).

[110] H. Bartl, Neues Jahrbuch fuer Mineralogie. Monatshefte 1970, 298-310, (1970).

[111] K. Stahl, G. Artioli, J.C. Hanson, Physics and Chemistry of Minerals 23, 328-336, (1996).

[112] J. Stolz, T. Armbruster, Neues Jahrbuch fuer Mineralogie. Monatshefte 1997, 131-144, (1997).

[113] H. Ghobarkar, O. Schäf, Mic. Mes. Materials 23, 55-60, 51998).

[114] J.G. Liou, Report (DOE/ER/12051-T1; Order No. DE85001293), 174 pp, (1984). From : Energy Res. Abstr. 1984, 9(23), Abstr. No. 48231.

[115] J.G. Liou, J. Petrol.12, 379-411, (1971).

[116] E. Galli, International Conference on Zeolites: Proceedings 205-213, (1980).

[117] G. Giuseppetti, F. Mazzi, C. Tadini, E. Galli, Neues Jahrbuch fuer Mineralogie. Monatshefte 1991, 307-314, (1991).

[118] H.W. Leimer, M. Slaughter, Zeitschrift fuer Kristallographie, Kristallgeometrie, Kristallphysik, Kristallchemie 130, 88-111, (1969).

[119] I.S. Kerr, D.J. Williams, Acta Crystallographica B 25, 1183-1190, (1969).

[120] A. Kvick, G. Artioli, J.V. Smith, Zeitschrift fuer Kristallographie 174, 265-281, (1986).

[121] I.S. Kerr, D.J. Williams, Zeitschrift fuer Kristallographie, Kristallgeometrie, Kristallphysik, Kristallchemie 125, 220-225, (1967).

[122] S. Khodabandeh, M.E. Davis, Microporous Materials 12(4-6), 347-359, (1997).

[123] D.B. Hawkins, Mater. Res. Bull. 2(11), 1021-1028, (1967).

[124] D.B. Hawkins, Mater. Res. Bull. 2(10), 951-958, (1967).

[125] R.M. Barrer, D.J. Marshall, Am. Mineralogist 50 484-9, (1965).

[126] K.F. Fischer, V. Schramm, Advances in Chemistry Series 98, 250-258, (1971).

[127] R. Rinaldi, G. Vezzalini, Studies in Surface Science and Catalysis 24, 481-492, (1985).

[128] G. Vezzalini, S. Quartieri, A. Alberti, Zeolites 13, 34-42, (1993).

[129] G. Artioli, R. Rinaldi, A. Kvick, J.V. Smith, Zeolites 6, 361-366, (1986).

[130] H. Ghobarkar, O. Schäf, Mat. Res. Bull. 34(4), 517-525, (1999).

[131] G.C.C Yang, T.-Y. Yang, Journal of Hazardous Materials 62(1), 75-89, (1998).

[132] B.R. Albert, A.K. Cheetham, C.J. Adams, Microporous and Mesoporous Materials 21(1-3), 127-132, (1998).

[133] S. Khodabandeh, M.E. Davis, Microporous Materials 9(3,4), 161-172, (1997).

[134] S. Hansen, U. Haakansson, A.R. Landa-Canovas, L. Faelth, Zeolites 13(4), 276-80, (1993).

[135] D.E.W. Vaughan, Mater. Res. Soc. Symp. Proc. 111(Microstruct. Prop. Catal.), 89-100, (1988).

[136] A. Erdem, L.B. Sand, Proc. 5th Int. Conf. Zeolites, 64-72, (1980).

[137] M. Sathupunya, E. Gulari, Journal of the European Ceramic Society 22(13), 2305-2314, (2002).

[138] G. Artioli, M. Marchi, Powder Diffraction 14, 190-194, (1998).

[139] G. Artioli, American Mineralogist, 77, 189-196, (1992).

[140] L. Schroepfer, W. Joswig, European Journal of Mineralogy 9, 53-65, (1997).

[141] G. Larsen, K.H. Plum, H. Foerster, Eur. J. Mineral. 3(6), 933-941, (1991).

[142] A. Alberti, G. Vezzalini, Acta Crystallographica B 35, 2866-2869, (1979).

[143] L.B. McCusker, C. Baerlocher, Zeitschrift fuer Kristallographie 171, (1985), 281-289.

[144] G. Artioli, H. Foy, Mineralogical Magazine 58, 615-620, (1994).

[145] R.J. Donahoe, J.G. Liou, S. Guldman, Clays and Clay Minerals 32(6), 433-443, (1984).

[146] R. Rinaldi, J.J. Pluth, J.V. Smith, Acta Crystallographica B 30, 2426-2433, (1974).

[147] H. Steinfink, Acta Crystallographica 15, 644-651, (1962).

[148] A.F. Gualtieri, Journal of Applied Crystallography 33, 267-278, (2000).

[149] A.F. Gualtieri, Acta Crystallographica B 56, 584-593, (2000).

[150] X. Querol, A. Alastuey, A. Lopez-Soler, F. Plana, J.M. Andres, R. Juan, P. Ferrer, C.R. Ruiz, Environmental Science and Technology 31(9), 2527-2533, (1997).

[151] P. Man, Park; C. Jyung, Clay Sci. 9(4), 231-239, (1995).

[152] D. Matulova, E. Klokocnikova, Acta Univ. Carol., Geol. (1-2), 193-202, (1994).

[153] N. Burriesci, M. Crisafulli, N. Giordano, P.L. Antonucci, Zeolites 6(2), 119-124, (1986).

[154] A. Cichocki, J. Grochowski, L. Lebioda, Krist. Tech. 14(1), 9-18, (1979).

[155] D.T. Hayhurst, L.B. Sand, ACS Symp. Ser. 40(Mol. Sieves-2, 4th Int. Conf.,), 219-232, (1977).

[156] W.C. Beard, Advan. Chem. Ser. 101(Molecular Sieve Zeolites-I), 237-249, (1971).

[157] G.H. Kuehl, Amer. Mineral. 54(11-12), 1607-1612, (1969).

[158] E.W. Albers, H.W. Burkhead, J.C.S. Shi, PCT Int. Patent 46 pp. WO 9609890 A1 19960404, (1996).

[159] J.E. McEvoy, James E. U.S. Patent 4 pp. US 3545921 19701208, (1970).

[160] J.E. McEvoy, G.A. Mills, U.S. Patent 6 pp. US 3532459 19701006, (1970).

[161] E. Michalko, U.S. Patent 6 pp. Continuation-in-part of U.S. 3359068. US 3386802 19680604, (1968).

[162] R. Rinaldi, J.J. Pluth, J.V. Smith, Acta Crystallographica B 30, 2426-2433, (1974).

[163] R. Sadanaga, F. Marumo, Y. Takeuchi, Acta Crystallographica 14, 1153-1163, (1961).

[164] E. Stuckenschmidt, H. Fuess, A. Kvick, European Journal of Mineralogy 2, 861-874, (1990).

[165] O. Chiyoda, M.E. Davis, U.S. Patent 14 pp., Cont.-in-part of U.S. 5,935,551. US 6187283 B1 20010213, (2001).

[166] M. Aoki, Ganseki Kobutsu Kosho Gakkaishi 73(6), 155-166, (1978).

[167] A.J. Perrotta, Am. Mineral. 61(5-6), 495-496, (1976).

[168] W.C. Beard, Advan. Chem. Ser. 101(Molecular Sieve Zeolites-I), 237-249, (1971).

[169] J.E. McEvoy, U.S. Patent 4 pp. US 3545921 19701208, (1970).

[170] A. Erdem, L.B. Sand, in : Proc. 5th Int. Conf. Zeolites, L.V.C. Rees, (Ed.), 64-72, (1980).

[171] R.M. Barrer, D.J. Marshall, J. Chem. Soc. 7, 2296-2305, (1964).

[172] E. Galli, Crystal Structure Communications 3, 339-344, (1974).

[173] E. Galli, Rendiconti della Societa Italiana di Mineralogia e Petrologia 31, 599-612, (1975).

[174] J.M. Newsam, R.H. Jarman, A.J. Jacobson, Materials Research Bulletin 20, 125-136, (1985).

[175] B.M. Skofteland, O.H. Ellestad, K.P. Lillerud, Microporous and Mesoporous Materials 43(1), 61-71, (2001).

[176] C. Colella, R. Aiello, V. Di Ludovico, Rend. Soc. Ital. Miner. Petrol. 33, 511-518, (1977).

[177] R.M. Barrer, D.E. Mainwaring, J. Chem. Soc. Dalton Trans. 2534-2546, (1972).

[178] H.-J. Lo, J.-C. Hu, S.-R. Song, L.-J. Lee, Journal of the Geological Society of China 42(1), 125-142, (1999).

[179] P.A. Barrett, S. Valencia, M.A. Camblor, Journal of Materials Chemistry 8(10), 2263-2268, (1998).

[180] J.C. Quirin, L. Yuen, S.I. Zones, Journal of Materials Chemistry 7(12), 2489-2494, (1997).

[181] A. Bienoik, K. Bornholdt, U. Brendel, W.H. Baur, Journal of Materials Chemistry 6(2), 271-275, (1996).

[182] X. Querol, A. Alastuey, J.L. Fernandez-Turiel, A. Lopez-Soler, Fuel 74(8), 1226-1231, (1995).

[183] J.M. Newsam, R.H. Jarman, A.J. Jacobson, Materials Research Bulletin 20, 125-136, (1985).

[184] E. Galli, Crystal Structure Communications 3, 339-344, (1974).

[185] E. Galli, Rendiconti della Societa Italiana di Mineralogia e Petrologia 31, 599-612, (1975).

[186] H. Ghobarkar, O. Schäf, U. Guth, High Pressure Research 20, 45-53, (2001).

[187] C. Bebon, D. Colson, B. Marrot, J.P. Klein, F. Di Renzo, Microporous and Mesoporous Materials 53(1-3), 13-20, (2002).

[188] A.M. Goossens, E.J.P. Feijen, G. Verhoeven, B.H. Wouters, P.J. Grobet, P.A. Jacobs, J.A. Martens, Microporous and Mesoporous Materials 35-36 555-572, (2000).

[189] R. Aiello, R.M. Barrer, J. Chem. Soc. Sect. A 1470-1475, (1970).

[190] F. Di Renzo, F. Fajula, F. Fitoussi, P. Schulz, Int. Patent WO 9746487 A2 19971211, (1997).

[191] B. De Witte, J. Patarin, J.L. Guth, T. Cholley, Microporous Materials 10(4-6), 247-257, (1997).

[192] F. Di Renzo, F. Fajula, F. Figueras, M.A. Nicolle, T. Des Courieres, Synth. Microporous Mater. 1 105-114, (1992).

[193] D.E.W. Vaughan, U.S. Patent US 5000932 A 19910319, (1991).

[194] T. Taga, S. Kasahara, Eur. Patent EP 273403 A1 19880706, (1988).

[195] G.V. Tsitsishvili, M.K. Charkviani, Studies in Surface Science and Catalysis 28 (New Dev. Zeolite Sci. Technol.), 161-167, (1986).

[196] E.K. Gordon, S. Samson, SW.B. Kamb, American Mineralogist 45, 79-79, (1960).

[197] E.K. Gordon, S. Samson, SW.B. Kamb, Science 154, 1004-1007, (1966).

[198] D.E.W. Vaughan, K.G. Strohmaier, Microporous and Mesoporous Materials 28(2), 233-239, (1999).

[199] K.G. Strohmaier, D.E.W. Vaughan, Eur. Patent EP 356082 A2 19900228, (1990).

[200] K.G. Strohmaier, D.E.W. Vaughan, Eur. Patent EP 213739 A2 19870311, (1987).

[201] Yu.A. Malinovskii, Kristallografiya 29, 426-430, (1984).

[202] E. Galli, E. Passaglia, P.F. Zanazzi, Neues Jahrbuch fuer Mineralogie. Monatshefte 1982, 145-155 , (1982).

[203] E. Galli, E. Passaglia, P.F. Zanazzi, Neues Jahrbuch fuer Mineralogie. Monatshefte 1978, 310-324, (1978).

[204] K. Fischer, Neues Jahrbuch fuer Mineralogie. Monatshefte 1966, 1-13, (1966).

[205] A.G. Vigdorchik, Yu.A. Malinovskii, Kristallografiya 31, 879-882, (1986).

[206] G. Vezzalini, S. Quartieri, E. Passaglia, Neues Jahrbuch fuer Mineralogie. Monatshefte 1990, 504-516, (1990).

[207] M. Sacerdoti, E. Passaglia, R. Carnevali, Zeolites 15, 276-281, (1995).

[208] H. Ghobarkar, O. Schäf, P. Knauth Angew. Chem. Int. Ed. 40(20), 3831-3833, (2001).

[209] O. Chiyoda, M.E. Davis, U.S. Patent 14 pp., Cont.-in-part of U.S. Ser. No. 836,966 US 6436364 B1 20020820, (2002).

[210] M.J. Rao, S.R. Krishna, K.P. Kumar, Indian Journal of Environmental Protection 20(4), 290-296, (2000).

[211] O. Chiyoda, M.E. Davis, Microporous and Mesoporous Materials 32(3), 257-264, (1999).

[212] X. Querol, F. Plana, Felicia; A. Alastuey, A. Lopez-Soler, Fuel 76(8), 793-799, (1997).

[213] M. Chatterjee, D. Ganguli, P. Saha, Prasenjit, Trans. Indian Ceram. Soc. 35(5), 99-105, (1976).

[214] D.C. Freeman Jr., D. Stamires, U.S. Patent 7 pp. DE 1196626 19650715, (1965).

[215] D.W. Breck, W.G. Eversole, R.M. Milton, J. Am. Chem. Soc. 78 2338-2339, (1956).

[216] R.M. Barrer, J.W. Baynham, F.W. Bultitude, W.M. Meier, J. Chem. Soc. (London) 195-208 (1959).

[217] A. Peyreron, J.L. Guth, R. Wey, C.R. Acad. Sci. Paris Ser. D 272, 181-184, (1971).

[218] R.M. Barrer, D.E. Mainwaring, J. Chem. Soc. Dalton Trans. 2534-2546, (1972).

[219] W.J. Mortier, J.J. Pluth, J.V. Smith, Materials Research Bulletin 12, 97-102, (1977).

[220] W.J. Mortier, J.J. Pluth, J.V. Smith, Acta Crystallographica 15, 835-835, (1962).

[221] W.J. Mortier, J.J. Pluth, J.V. Smith, Acta Crystallographica 16, 45-45, (1963).

[222] W.J. Mortier, J.J. Pluth, J.V. Smith, Acta Crystallographica 17, 384-384, (1964).

[223] W.J. Mortier, G.S.D. King, L. Sengler, Journal of Physical Chemistry 83, 2263-2266, (1979).

[224] M. Calligaris, G. Nardin, Zeolites 2, 200-204, (1982).

[225] A. Alberti, E. Galli, G. Vezzalini, E. Passaglia, P.F. Zanazzi, Zeolites 2, 303-309, (1982).

[226] M. Calligaris, A. Mezzetti, G. Nardin, L. Randaccio, Zeolites 4, 323-328, (1984).

[227] M. Calligaris, G. Nardin, L. Randaccio, Zeolites 4, 251-254, (1984).

[228] R.M. Barrer, J.W. Baynham, J. Chem. Soc. (London) 2882-2891, (1956).

[229] K. Tomita, H. Yamashita, N. Oba, J. Jap. Assoc. Miner. Petr. Econ. Geol. 62, 80-89, (1969).

[230] A. Alberti, C. Colella, G. Oggiano, M. Pansini, G. Vezzalini, Materials Engineering 5(2), 145-158(1994).

[231] P.K. Dutta, R. Asiaie, Synth. Microporous Mater. 1 522-536, (1992).

[232] C.G. Coe, T.R. Gaffney, R. Srinivasan, T. Naheiri, Process Technology Proceedings 11(Separation Technology), 267-279, (1994).

[233] A. Cichocki, Zeolites 11(8), 758-766, (1991).

[234] M. Chatterjee, D. Ganguli, P. Saha, Trans. Indian Ceram. Soc. 35(5), 99-105, (1976).

[235] K. Tomita, H. Yamashita, N. Oba, Ganseki Kobutsu Kosho Gakkaishi 62(2), 80-89, (1969).

[236] C. Amrhein, G.H. Haghnia, T.S. Kim, P.A. Mosher, R.C. Gagajena, T. Amanios, L. de la Torre, Environmental Science and Technology 30(3), 735-742, (1996).

[237] E. Piera, M.A. Salomon, J. Coronas, M. Menendez, J. Santamaria, Journal of Membrane Science 149(1), 99-114, (1998).

[238] H. Lee, P.K. Dutta, Microporous and Mesoporous Materials 38(2-3), 151-159, (2000).

[239] U. Mueller, F. Hill, N. Rieber, German Patent (Offen.) 6 pp. DE 19939416 A1 20010222, (2001).

[240] E. Tillmanns, R.X. Fischer, H. Baur, Neues Jahrbuch fuer Mineralogie. Monatshefte 1984, 547-558, (1984).

[241] G. Vezzalini, S. Quartieri, E. Galli, Zeolites 19, 75-79, (1997).

[242] S. Merlino, E. Galli, A. Alberti, Tschermaks Mineralogische und Petrographische Mitteilungen 22, 117-129, (1975).

[243] M. Sacerdoti, Neues Jahrbuch fuer Mineralogie. Monatshefte 1996, 114-124, (1996).

[244] C.V. Tuoto, A. Regina, J.B. Nagy, A. Nastro, Microporous and Mesoporous Materials 20(4-6), 247-257, (1998).

[245] C.V. Tuoto, J.B. Nagy, A. Nastro, Studies in Surface Science and Catalysis 97(Zeolites: A Refined Tool for Designing Catalytic Sites), 551-556, (1995).

[246] P. Caullet, PL. Delmotte, A.C. Faust, J.L. Guth, Zeolites 15(2), 139-147, (1995).

[247] U. Wirsching, Clays Clay Miner. 29(3), 171-183, (1981).

[248] J.L. Schlenker, J.J. Pluth, JJ.V. Smith, Acta Crystallographica B 33, 3265-3268, (1977).

[249] A. Kawahara, H. Curien, Bulletin de la Societe Francaise de Mineralogie et de Cristallographie 92, 250-256, (1969).

[250] L.W. Staples, J.A. Gard, Mineralogical Magazine and Journal of the Mineralogical Society 32, 261-281, (1959).

[251] A. Alberti, A. Martucci, E. Galli, G. Vezzalini, Zeolites 19, 349-352, (1997).

[252] A. Gualtieri, G. Artioli, E. Passaglia, S. Bigi, A. Viani, J.C. Hanson, American Mineralogist 83, 590-606, (1998).

[253] W. Inaoka, S. Kasahara, T. Fukushima, K. Igawa, Studies in Surface Science and Catalysis 60(Chem. Microporous Cryst.), 37-42, (1991).

[254] L.M. Occelli, R.A. Innes, S.S. Pollack, J.V. Sanders, Zeolites 7(3), 265-271, (1987).

[255] K.P. Lillerud, J.H. Raeder, Zeolites 6(6), 474-483, (1986).

[256] S. Ueda, M. Nishimura, M. Koizumi, Studies in Surface Science and Catalysis 24(Zeolites: Synth., Struct., Technol. Appl.), 105-110, (1985).

[257] F.G. Van den Berg, J.H.E. Glezer, Brit. UK Pat. Appl. 4 pp. GB 2128972 A1 19840510, (1984).

[258] H.J. Lo, Proceedings of the National Science Council, Republic of China, Part A: Applied Sciences 7(2), 69-74, (1983).

[259] H.E. Robson, French Patent 6 pp. FR 1516447 19680308, (1968).

[260] A. Cichocki, P. Koscielniak, Microporous and Mesoporous Materials 29(3), 369-382, (1999).

[261] J.E. Gilbert, A. Mosset, European Journal of Solid State and Inorganic Chemistry 35(6-7), 447-458, (1998).

[262] A. Cichocki, P. Koscielniak, M. Michalik, M. Bus, Zeolites 18(1), 25-32, (1997).

[263] S. Krais, Petroleum and Coal 38(1), 57-64, (1996).

[264] S. Zainea, C. Nichita, G. Gheorghe, C. Nichita, Proceedings of the 13th International Congress on Electron Microscopy, Paris, July 17-22, 2B 1299-1300, (1994).

[265] S. Yang, N.P. Evmiridis, Microporous Materials 6(1), 19-26, (1996).

[266] A. Gualtieri, G. Artioli, E. Passaglia, S. Bigi, A. Viani, J.C. Hanson, American Mineralogist 83, 590-606, (1998).

[267] J.A. Gard, J.M. Tait, Acta Crystallographica B 28, 825-834, (1972).

[268] W.J. Mortier, J.J. Pluth, J.V. Smith, Zeitschrift fuer Kristallographie, Kristallgeometrie, Kristallphysik, Kristallchemie 144, 32-41, (1976).

[269] W.J. Mortier, J.J. Pluth, J.V. Smith, Zeitschrift fuer Kristallographie, Kristallgeometrie, Kristallphysik, Kristallchemie 143, 319-332, (1976).

[270] A. Alberti, G. Cruciani, E. Galli, G. Vezzalini, Zeolites 17, 457-461, (1996).

[271] H. Ghobarkar, O. Schäf, Cryst. Res. Technol. **31**(3), K29-K31, (1996).

[272] M.G. Howden, Zeolites 7(3), 255-259, (1987).

[273] W.H. Flank, U.S. Patent 4 pp. US 3758539 19730911, (1973).

[274] J .P.Verduijn, Int. Patent Appl. 34 pp. WO 9214680 A1 19920903, (1992).

[275] J. Groeb, F. Fetting, Chemie Ingenieur Technik 61(10), 817-818, (1989).

[276] M.G. Howden, Zeolites 7(3), 260-264, (1987).

[277] L. Moudafi, P. Massiani, F. Fajula, F. Figueras, Zeolites 7(1), 63-66, (1987).

[278] G. Ferre, P. Dufresne, C. Marcilly, Ger. Patent Offen. 23 pp. DE 3530947 A1 19860313, (1986).

[279] L. Moudafi, R. Dutartre, F. Fajula, F. Figueras, Applied Catalysis 20(1-2), 189-203, (1986).

[280] P. Rouet, J.L. Guth, P. Dufresne, R. Wey, Ger. Patent Offen. 17 pp. DE 3520508 A1 19851212, (1985).

[281] F. Fajula, F. Figueras, L. Moudafi, Eur. Pat. Appl. 24 pp. EP 118382 A1 19840912, (1984).

[282] L.D. Rollmann, E.W. Valyocsik, Inorganic Syntheses 22 61-68, (1983).

[283] L.B. Sand, U.S. Patent 11 pp. US 4093699 19780606, (1978).

[284] F. Gao, G. Zhu, X. Li, S. Qiu, B. Wei, C. Shao, O. Terasaki, Studies in Surface Science and Catalysis 135(Zeolites and Mesoporous Materials at the Dawn of the 21st Century), 519-526, (2001).

[285] S.S. Khvoshchev, M.A. Shubaeva, I.V. Karetina, Yu.V. Shapoval, Proceedings of the 12th International Zeolite Conference, Baltimore, M.M.J. Treacy (ed.) 3 1961-1968, (1999).

[286] S. Yang, A.G. Vlessidis, N.P. Evmiridis, Microporous Materials 9(5,6), 273-286, (1997).

[287] J.P. Verduijn Int. Patent Appl. 43 pp. WO 9703020 A1 19970130, (1997).

[288] M.L. Occelli, A.J. Perrotta, ACS Symposium Series 218(Intrazeolite Chem.), 21-39, (1983).

[289] E.L. Wu, T.E. Whyte Jr, M.K. Rubin, P.B. Venuto, J. Catal. 33(3), 414-419, (1974).

[290] E.E. Jenkins, U.S. Patent 7 pp. US 3578398 19710511, (1971).

[291] G. Bergerhoff, W.H. Baur, W. Nowacki, Neues Jahrbuch fuer Mineralogie. Monatshefte 1958, 193-200, (1958).

[292] W.H. Baur, American Mineralogist 49, 697-704, (1964).

[293] Yu.F. Shepelev, A.A. Anderson, Yu.I. Smolin, Kristallografiya 33, 359-364, (1988).

[294] D.H. Olson, E. Dempsey, Journal of Catalysis 13, 221-231, (1969).

[295] D.H. Olson, Journal of Physical Chemistry 72, 4366-4373, (1968).

[296] M.J. Bennett, J.V. Smith, Materials Research Bulletin 3, 633-642, (1968).

[297] J.J. Pluth, J.V. Smith, Materials Research Bulletin 7, 1311-1322, (1972).

[298] P.K. Maher, F.D. Hunter, J. Scherzer, Advances in Chemistry Series 101, 266-278, (1971).

[299] S. Rayalu, N.K. Labhasetwar, P. Khanna, Brit. Patent Appl. 34 pp. GB 2339774 A1 20000209, (2000).

[300] D.E.W.Vaughan, K.G. Strohmaier, Studies in Surface Science and Catalysis 28(New Dev. Zeolite Sci. Technol.), 207-213, (1986).

[301] S. Kasahara, K. Itabashi, K. Igawa, Studies in Surface Science and Catalysis 28(New Dev. Zeolite Sci. Technol.), 185-192, (1986).

[302] G.C. Edwards, D.E.W. Vaughan, E.W. Albers, Can. Patent 17 pp. CA 1073431 19800311, (1980).

[303] R. Aiello, C. Colella, Ann. Chim. (Rome) 61(2), 122-130, (1971).

[304] H.E. Robson, U.S. Patent 7 pp. US 3343913 19670926, (1967).

[305] W.L. Haden Jr., F.J. Dzierzanowski, U.S. Patent 5 pp. US 3338672 19670829, (1967).

[306] Mobil Oil Corp., Brit. Patent 17 pp. GB 1070906 19670607, (1967).

[307] Esso Research and Engineering Co., Neth. Patent Appl. 25 pp. NL 6609757 19670116, (1967).

[308] D.W. Breck, W.G. Eversole, R.M. Milton, J. Am. Chem. Soc. 78 2338-2339, (1956).

[309] A. Elo Jr., F.R. Broersma, U.S. Patent 2 pp. US 3676063 19720711, (1972).

[310] C.H. Elliott Jr., C.V. McDaniel, U.S. Patent 4 pp. US 3639099 19720201, (1972).

[311] G. Heinze, F. Schwochow, H. Weber, Brit. Patent 5 pp. GB 1237551 19710630, (1971).

[312] C. Colella, R. Aiello, Annali di Chimica (Rome) 60(8-9), 587-596, (1970).

[313] P.K. Maher, J. Scherzer, U.S. Patent 3 pp. US 3516786 19700623, (1970).

[314] S.G. Hindin, J.C. Dettling, Brit. Patent 6 pp. GB 1167775 19691022, (1969).

[315] G.C. Edwards, D.E.W. Vaughan, E.W. Albers, U.S. Patent 9 pp. US 4175059 19791120, (1979).

[316] E.S. Rogers, Ger. Patent Offen. 16 pp. DE 2807660 19780831, (1978).

[317] D.E.W. Vaughan, G.C. Edwards, M.G. Barrett, Ger. Patent Offen. 24 pp. DE 2703264 19770804, (1977).

[318] M. Chatterjee, D. Ganguli, P. Saha, Transactions of the Indian Ceramic Society 34(5), 87-93, (1975).

[319] H. Kacirek, H. Lechert, Ger. Patent Offen. 6 pp. DE 2447206 19760408, (1976).

[320] D.E.W. Vaughan, U.S. Patent 9 pp. US 4714601 A 19871222, (1987).

[321] P.I. Reid, Eur. Patent Appl. 6 pp. EP 209332 A2 19870121, (1987).

[322] M.S. Joshi, V.V. Joshi, Bulletin of Materials Science 7(5), 475-481, (1985).

[323] N. Burriesci, M. Crisafulli, N. Giordano, P.L. Antonucci, Zeolites 6(2), 119-124, (1986).

[324] J. Arika, M. Aimoto, H. Miyazaki, Eur. Patent Appl. 34 pp. EP 128766 A2 19841219, (1984).

[325] R.N. Sanders, U.S. Patent 9 pp. US 4400366 A 19830823, (1983).

[326] F.J. Fiedler, H.H. Lohse, K. Schuermann, Neues Jahrbuch fuer Mineralogie, Monatshefte (8), 358-364, (1983).

[327] P. Bosch, L. Ortiz, I. Schifter, Industrial & Engineering Chemistry Product Research and Development 22(3), 401-406, (1983).

[328] S.L. Burkett, M.E. Davis, Microporous Materials 1(4), 265-282, (1993).

[329] F. Dougnier, J.L. Guth, J. Patarin, D. Anglerot, Eur. Patent Appl. 12 pp. EP 530104 A1 19930303, (1993).

[330] T. Chatelain, J. Patarin, E. Brendle, F. Dougnier, J.L. Guth, P. Schulz, Studies in Surface Science and Catalysis 105A(Progress in Zeolite and Microporous Materials, pt. A), 173-180, (1997).

[331] P.C. Hu, E.W. Liimatta, Int. Patent Appl. 14 pp. WO 9602462 A1 19960201, (1996).

[332] T. Lindner, H. Lechert, Zeolites 16(2/3), 196-206, (1996).

[333] A.J. Caglione, T.R. Cannan, N. Greenlay, R.J. Hinchey, U.S. Patent 5 pp. US 5366720 A 19941122, (1994).

[334] G. Zhu, S. Qiu, J. Yu, Y. Sakamoto, F. Xiao, R. Xu, O. Terasaki, Chemistry of Materials 10(6), 1483-1486, (1998).

[335] S.J. Miller, U.S. Patent 9 pp. US 5716593 A 19980210, (1998).

[336] C. Bebon, D. Colson, B. Marrot, J.P. Klein, Jean F. Di Renzo, Microporous and Mesoporous Materials 53(1-3), 13-20, (2002).

[337] Q. Li, D. Creaser, J. Sterte, Studies in Surface Science and Catalysis 135(Zeolites and Mesoporous Materials at the Dawn of the 21st Century), 151-158, (2001).

[338] M.P. Kuvettu, M. Santra, V. Krishnan, S.K. Ray, J. Christopher, G.S. Mishra, R.M. Thakur, S. Makhija, S. Ghosh, U.S. Patent 6 pp. US 6284218 B1 20010904, (2001).

[339] R.C. Rouse, D.R. Peacor, American Mineralogist 71, 1494-1501, (1986).

[340] N. Almora-Barrios, A.R. Ruiz-Salvador, A. Gomez, M. Mistry, D.W. Lewis, Chemical Communications (Cambridge, U.K.) (6), 531-532, (2001).

[341] K. Shiokawa, M. Ito, K. Itabashi, Zeolites 9, 170-176, (1989).

[342] J.L. Schlenker, J.J. Pluth, J.V. Smith, Materials Research Bulletin 13, 169-174, (1978).

[343] J.L. Schlenker, J.J. Pluth, J.V. Smith, Materials Research Bulletin 14, 751-758, (1979).

[344] J.L. Schlenker, J.J. Pluth, J.V. Smith, Materials Research Bulletin 14, 961-966, (1979).

[345] W.M. Meier, Zeitschrift fuer Kristallographie, Kristallgeometrie, Kristallphysik, Kristallchemie 115, 439-450, (1961).

[346] M. Ito, Y. Saito, Bulletin of the Chemical Society of Japan 58, 3035-3036, (1985).

[347] A. Alberti, P. Davoli, G. Vezzalini, Zeitschrift fuer Kristallographie, Kristallgeometrie, Kristallphysik, Kristallchemie 175, 249-256, (1986).

[348] J. Elsen, G.S.D. King, W.J. Mottier, Journal of Physical Chemistry 91, 5800-5805, (1987).

[349] H. Ghobarkar, O. Schäf, U. Guth, High Pressure Research 20, 45-53, (2001).

[350] S. Komarneni, Zeolites 6(2), 95-98, (1986).

[351] R.M. Barrer, J. Chem. Soc. (London) 2158-2163, (1948).

[352] L.B. Sand in : Molecular Sieves. Society of Chemical Industry, London, 71-77, (1968).

[353] E. Piera, M.A. Salomon, J. Coronas, M. Menendez, J. Santamaria, Journal of Membrane Science 149(1), 99-114, (1998).

[354] J. Warzywoda, A.G. Dixon, R.W. Thompson, A. Sacco Jr., Journal of Materials Chemistry 5(7), 1019-1025, (1995).

[355] G.J. Kim, W.S. Ahn, Zeolites 11(7), 745-750, (1991).

[356] P. Bodart, J.B. Nagy, E.G. Derouane, Z. Gabelica, Studies in Surface Science and Catalysis 18(Struct. React. Modif. Zeolites), 125-132, (1984).

[357] K.P. Bajpai, M.S. Rao, K.V.G.K. Gokhale, Industrial & Engineering Chemistry Product Research and Development 20(4), 721-726, (1981).

[358] D. Domine, J. Quobex, Mol. Sieves, Conf. Paper 78-84, (1968).

[359] R.W. Aitken, I.M. Keen, S. African Patent 12 pp. ZA 6705138 19680124, (1968).

[360] L.J. Reid Jr., U.S. Patent 2 pp. US 3334964 19670808, (1967).

[361] A.J. Ellis, Geochim. et Cosmochim. Acta 19 145-146, (1960).

[362] F. Wolf, M. Stirn, Silikattechnik 23(7), 227-229, (1972).

[363] F. Wolf, J. Renning, Tonindustrie-Zeitung und Keramische Rundschau 97(11), 293-298, (1973).

[364] O.J. Whittemore Jr., American Mineralogist 57(7-8), 1146-1151, (1972).

[365] M.L. Sand, W.S. Coblenz, L.B. Sand, Advances in Chemistry Series 101(Molecular Sieve Zeolites-I), 127-134, (1971).

[366] D.G. Stewart, W.J. Ball, Eur. Patent Appl. 17 pp. EP 14023 19800806, (1980).

[367] D.B. Shukla, P.M. Oza, V.P. Pandya, Research and Industry 25(1), 8-10, (1980).

[368] L.D. Rollmann, E.W. Valyocsik, U.S. Patent 6 pp. US 4205052 19800527, (1980).

[369] R.Y. Saleh, P.B. Koradia, Ger. Patent Offen. 7 pp. DE 2934382 19800327, (1980).

[370] M.L. Occelli, S.S. Pollack, J.V. Sanders, Studies in Surface Science and Catalysis 37(Innovation Zeolite Mater. Sci.), 45-55, (1988).

[371] G. Bellussi, G. Perego, A. Carati, U. Cornaro, V. Fattore, Studies in Surface Science and Catalysis 37(Innovation Zeolite Mater. Sci.), 37-44, (1988).

[372] M.S. Joshi, K.M. Prabhu, Bulletin of Materials Science 9(1), 7-11, (1987).

[373] M.J. Desmond, F.A. Pesa, J.K. Currie, U.S. Patent 8 pp. US 4585640 A 19860429, (1986).

[374] S. Komarneni, Zeolites 6(2), 95-98, (1986).

[375] P.K. Bajpai, Zeolites 6(1), 2-8, (1986).

[376] K. Itabashi, T. Fukushima, K. Igawa, Zeolites 6(1), 30-34, (1986).

[377] C.H. Chi, L.B. Sand, Zeolites 5(5), 309-312, (1985).

[378] A. Meier, W. Lukas, M. Wilhelmy, P. Hantschel, Int. Patent Appl. 13 pp. WO 9114652 A1 19911003, (1991).

[379] G.J. Kim, W.S. Ahn, Zeolites 11(7), 745-750, (1991).

[380] D.R. Simpson, U.S. Patent 4 pp. Cont.-in-part of U.S. Ser. No. 27,792, abandoned. US 4935217 A 19900619, (1990).

[381] A.J. Chandwakdar, P. Ratnasamy, Indian Patent 18 pp. IN 172653 A 19931030, (1993).

[382] Y. Sun, S. Qui, T. Song, W. Pang, Y. Yue, Journal of the Chemical Society, Chemical Communications (13), 1048-1050, (1993).

[383] S.D. Hellring, R.F. Striebel, U.S. Patent 10 pp. US 5219547 A 19930615, (1993).

[384] H. Du, S. Qiu, W. Pang, Chemical Research in Chinese Universities 11(4), 364-366, (1995).

[385] Y. Sun, T. Song, S. Qiu, W. Pang, J. Shen, D. Jiang, Y. Yue, Zeolites 15(8), 745-753, (1995).

[386] C.S.H. Chen, U.S. Patent 19 pp. US 5573746 A 19961112, (1996).

[387] P. De Luca, F. Crea, R. Aiello, A. Fonseca, J.B. Nagy, Studies in Surface Science and Catalysis 105A(Progress in Zeolite and Microporous Materials, pt. A), 325-332, (1997).

[388] M.A. Zanjanchi, G. Vaghar-Lahijani, M. Mohammadian, Journal of Sciences (Iran) 9(2), 156-162, (1998).

[389] R. Millini, L. Carluccio, F. Frigerio, W. O'Neil Parker Jr., G. Bellussi, Microporous and Mesoporous Materials 24(4-6), 199-211, (1998).

[390] M.L. Pavlov, R.A. Makhamatkhanov, B.I. Kutepov, F.Kh. Kudasheva, Yu.A. Lebedev, Russian Journal of Applied Chemistry 75(8), 1360-1361, (2002).

[391] G. Gottardi, W.M. Meier, Zeitschrift fuer Kristallographie, Kristallgeometrie, Kristallphysik, Kristallchemie 119, 53-64, (1963).

[392] G. Vezzalini, Zeitschrift fuer Kristallographie 166, 63-71, (1984).

[393] G. Vezzalini, Zeitschrift fuer Kristallographie, Kristallgeometrie, Kristallphysik, Kristallchemie 119, 53-64, (1963).

[394] G.M. Alieva, G.F. Mamedova, Z.B. Abdullaev, D.M. Ganbarov, Azerbaidzhanskii Khimicheskii Zhurnal (1), 86-88, (2001).

[395] A.J. Perrotta, Mineralogical Magazine and Journal of the Mineralogical Society 36, 480-490, (1967).

[396] M. Slaughter, W.T. Kane, Zeitschrift fuer Kristallographie, Kristallgeometrie, Kristallphysik, Kristallchemie 130, 68-87, (1969).

[397] A. Alberti, E. Galli, G. Vezzalini, Zeitschrift fuer Kristallographie 173, 257-265, (1985).

[398] P. Yang, T. Armbruster, European Journal of Mineralogy 8, 263-271, (1996).

[399] H. Ghobarkar, Crystal Res. and Technol. 19(12), 1571-1573, (1984).

[400] M.E. Davis, S. Khodabandeh, Int. Patent Appl. 29 pp. WO 9718163 A1 19970522, (1997).

[401] H.J. Lo, Zhongguo Dizhi Xuehui Huikan 24 9-20, (1981).

[402] I.J. Pichering, P.J. Maddox, J.M. Thomas, A.K. Cheetham, Journal of Catalysis 119, 261-265, (1989).

[403] A. Alberti, C. Sabelli, Zeitschrift fuer Kristallographie 178, 249-256, (1987).

[404] H. Gies, R.P. Gunawardane, Zeolites 7, 442-445, (1987).

[405] R.E. Morris, S.J. Weigel, N.J. Henson, L.M. Bull, M.T. Janicke, B.F. Chmelka, A.K. Cheetham, Journal of the American Chemical Society 116, 11849-11855, (1994).

[406] R. Gramlich-Meier, W.M. Meier, B.K. Smith, Zeitschrift fuer Kristallographie 169, 201-210, (1984).

[407] R. Gramlich-Meier, W.M. Meier, B.K. Smith, Acta Crystallographica 21, 983-990, (1966).

[408] P.A. Vaughan, Acta Crystallographica 21, 983-990, (1966).

[409] S.J. Weigel, J.-C. Gabriel, E. Gutierrez-Puebla, A. Monge Bravo, N.J. Henson, L.M. Bull, A.K. Cheetham, Journal of the American Chemical Society 118, 2427-2435, (1996).

[410] S. Popescu, R. Russu, C. Nichita, A. Visan in : B. Jouffrey, C. Colliex (eds.), Proceedings of the International Congress on Electron Microscopy, 2B, 811-812, (1994).

[411] L. Schreyeck, P. Caullet, J.-C. Mougenel, J.L. Guth, B. Marler, Journal of the Chemical Society, Chemical Communications 21, 2187-2188, (1995).

[412] N.R. Forber, L.V.C. Rees, Zeolites 15(5), 444-451, (1995).

[413] S. Nadimi, S. Oliver, A. Kuperman, A. Lough, G.A. Ozin, J.M. Garces, M.M. Olken, P. Rudolf, Studies in Surface Science and Catalysis 84(Zeolites and Related Microporous Materials, Pt. A), 93-100, (1994).

[414] N. Kanno, M. Miyake, M. Sato, Zeolites 14(8), 625-628, (1994).

[415] C.D. Chang, S.D. Hellring, U.S. Patent 6 pp. US 5288475 A 19940222, (1994).

[416] J. Li, G. Liu, Ju. Li, Catalysis Letters 20(3-4), 345-348, (1993).

[417] M. Matsukata, N. Nishiyama, K. Ueyama, Microporous Materials 1(3), 219-222, (1993).

[418] R.M. Barrer, D.J. Marshall, J. Chem. Soc. (London), 485-497, (1964).

[419] S.D. Hellring, C.D. Chang, J.D. Lutner, U.S. Patent 11 pp. US 5190736 A 19930302, (1993).

[420] A. Culfaz, A.K. Yilmaz, Crystal Research and Technology 20(1), 11-19, (1985).

[421] L. Marosi, M. Schwarzmann, J. Stabenow, Eur. Patent Appl. 8 pp. EP 49386 A1 19820414, (1982).

[422] D.E.W. Vaughan, G.C. Edwards, U.S. Patent 8 pp. US 4088739 19780509, (1978).

[423] W.E. Cormier, L.B. Sand, U.S. Patent 6 pp. US 4017590 19770412, (1977).

[424] W.E. Cormier, L.B. Sand, B. Leonard, Am. Mineral. 61(11-12), 1259-1266, (1976).

[425] R.K. Ahedi, A.N. Kotasthane, Journal of Porous Materials 4(3), 171-179, (1997).

[426] G.M. Pasquale, B.D. Murray, Int. Patent Appl. 37 pp. WO 9640587 A1 19961219, (1996).

[427] M. Matsukata, N. Nishiyama, K. Ueyama, Microporous Materials 7(2-3), 109-117, (1996).

[428] T.J. Kim, W.S. Ahn, S.B. Hong, Microporous Materials 7(1), 35-40, (1996).

[429] R.A. Rakoczy, M. Breuninger, M. Hunger, Y. Traa, J. Weitkamp, Chemical Engineering & Technology 25(3), 273-275, (2002).

[430] G.-Q. Guo, Y.-J. Sun, Y.-C. Long, Studies in Surface Science and Catalysis 135(Zeolites and Mesoporous Materials at the Dawn of the 21st Century), 527-533, (2001).

[431] R.B. Khomane, B.D. Kulkarni, R.K. Ahedi, Journal of Colloid and Interface Science 236(2), 208-213, (2001).

[432] G. Pal-Borbely, H.K. Beyer, Y. Kiyozumi, F. Mizukami, Microporous and Mesoporous Materials 22(1-3), 57-68, (1998).

[433] V. Kocman, R.I. Gait, J.C. .;Rucklidge, American Mineralogist 59, 71-78, (1974).

[434] G. Bissert, F. Liebau, Neues Jahrbuch fuer Mineralogie. Monatshefte 1986, 241-252, (1986).

[435] K. Stahl, A. Kvick, S. Ghose, Zeolites 9, 303-311, (1989).

[436] S. Quartieri, A. Sani, G. Vezzalini, E. Galli, E. Fois, A. Gamba, G. Tabacchi, Microporous and Mesoporous Materials 30, 77-87, (1999).

[437] D.J. Drysdale, Amer. Mineral. 56(9-10), 1718-1723, (1971).

[438] M.K. Rubin, U.S. Patent 7 pp. Cont.-in-part of U.S. Ser. No. 17,288, abandoned. US 4820502 A 19890411, (1989).

[439] A.B. Merkle, M. Slaughter, American Mineralogist 53, 1120-1138, (1968).

[440] A. Alberti, Tschermaks Mineralogische und Petrographische Mitteilungen 18, 129-146, (1972).

[441] A.B. Merkle, M. Slaughter, American Mineralogist 52, 273-276, (1967).

[442] N. Bresciani-Pahor, M. Calligaris, G. Nardin, L. Randaccio, E. Russo, P. Comin-Chiaramonti, Journal of the Chemical Society. Dalton Transactions, Inorganic Chemistry 1980, 1511-1514, (1980).

[443] E. Galli, G. Gottardi, H. Mayer, A. Preisinger, E. Passaglia, Acta Crystallographica B 39, 189-197, (1983).

[444] M.E. Gunter, T. Armbruster, T. Kohler, C.R. Knowles, American Mineralogist 79, 675-682, (1994).

[445] T.W. Hambley, J.C. Taylor, Journal of Solid State Chemistry 54, 1-9, (1984).

[446] P. Yang, T. Armbruster, Journal of Solid State Chemistry 123, 140-149, (1996).

[447] K. Sugiyama, Y. Takeuchi, Studies in Surface Science and Catalysis 28, 449-456, (1986).

[448] T. Wuest, J. Stolz, T. Armbruster, American Mineralogist 84, 1126-1134, (1999).

[449] P. Yang, T. Armbruster, European Journal of Mineralogy 10, 461-471, (1998).

[450] A. Sani, G. Vezzalini, P. Ciambelli, M.T. Rapacciuolo, Microporous and Mesoporous Materials 31, 263-270, (1999).

[451] N. Bresciani-Pahor, M. Calligaris, G. Nardin, L. Randaccio, Journal of the Chemical Society. Dalton Transactions, Inorganic Chemistry 1981, 2288-2291, (1981).

[452] H.Ghobarkar, O. Schäf, B. Paz, P. Knauth: J. Solid State Chem., 173, 27-31, (2003).

[453] D. Zhao, K. Cleare, C. Oliver, C. Ingram, D. Cook, R. Szostak, L. Kevan, Microporous and Mesoporous Materials 21(4-6), 371-379, (1998).

[454] D. Zhao, R. Szostak, L. Kevan, Journal of Materials Chemistry 8(1), 233-239, (1998).

[455] D. Zhao, L. Kevan, R. Szostak, Zeolites 19(5/6), 366-369, (1997).

[456] S. Khodabandeh, M.E. Davis, Microporous Materials 9(3,4), 149-160, (1997).

[457] L.L. Ames, American Mineralogist 48, 1372-1381, (1963).

[458] J.R. Smyth, A.T. Spaid, D.L. Bish, American Mineralogist 75, 522-528, (1990).

[459] A. Alberti, Tschermaks Mineralogische und Petrographische Mitteilungen 22, 25-37, (1975).

[460] T. Armbruster, M.E. Gunter, American Mineralogist 76, 1872-1883, (1991).

[461] K. Sugiyama, Y. Takeuchi, Studies in Surface Science and Catalysis 28, 449-456, (1986).

[462] P. Cappelletti, A. Langella, G. Cruciani, European Journal of Mineralogy 11, 1051-1060, (1999).

[463] O.E. Petrov, L.D. Filizova, G.N. Kirov, Dokl. Bulg. Akad. Nauk. 44, 77-80, (1991).

[464] T. Armbruster, American Mineralogist 78, 260-264, (1993).

[465] Y. Goto, American Mineralogist 62(3-4), 330-332, (1977).

[466] C.H. Chi, L.B. Sand, Nature 304(5923), 255-257, (1983).

[467] G. Kirov, V. Pechigargov, E. Landzheva, Chemical Geology 26(1-2), 17-28, (1979).

[468] D.B. Hawkins, R.A. Sheppard, A.J. Gude III., in : L.B. Sand, F.A. Mumpton (eds.), Nat. Zeolites: Occurrence, Prop., Use, Sel. Pap. Zeolite, Int. Conf. Proc. 337-343, (1978).

[469] S. Satokawa, K. Itabashi, Eur. Pat. Appl. 10 pp. EP 681991 A1 19951115, (1995).

[470] G. Seo, M.-W. Kim, J.-H. Kim, B.J. Ahn, S.B. Hong, Y.S. Uh, Catalysis Letters 55(2), 105-112, (1998).

[471] M. Kato, S. Satokawa, K. Itabashi, Studies in Surface Science and Catalysis 105A 229-235, (1997).

[472] D. Zhao, R. Szostak, L. Kevan, Journal of Materials Chemistry 8(1), 233-239, (1998).

[473] C.D. Williams, Chemical Communications (Cambridge) (21), 2113-2114, (1997).

[474] S. Satokawa, K. Itabashi, Microporous Materials 8(1,2), 49-55, (1997).

[475] S. Quartieri, G. Vezzalini, Zeolites 7, 163-170, (1987).

[476] E. Galli, Acta Crystallographica B 27, 833-841, (1971).

[477] M. Slaughter, American Mineralogist 55, 387-397, (1970).

[478] E. Galli, G. Gottardi, Mineralogica et Petrographica Acta 12, 1-10, (1966).

[479] J.R. Pearce, W.J. Mortier, G.S.D. King, J.J. Pluth, I.M. Steele, J.V. Smith, Proc. International Conference on Zeolites 261-268, (1980).

[480] W.J. Mortier, American Mineralogist 68, 414-419, (1983).

[481] M. Akizuki, Y. Kudoh, Y. Satoh, European Journal of Mineralogy 5, 839-843, (1993).

[482] H. Ghobarkar, O. Schäf, Journal of Physics D: Applied Physics 31, 3172-3176, (1998).

[483] H. Ghobarkar, O. Schäf and U. Guth, Journal of Solid State Chemistry 142, 451-454, (1999).

[484] C.D.R. De Souza, L.R.D. Da Silva, Anais da Associacao Brasileira de Quimica 49(3), 102-106, (2000).

[485] T. Tsanov, Y. Pavlova, P. Karamisheva, Bulg. Godishnik na Visshiya 26(1), 19-27, (1980).

[486] E. Passaglia, M. Sacerdoti, Bulletin de Mineralogie 105, 338-342, (1982).

[487] E. Galli, A. Alberti, Bulletin de la Societe Francaise de Mineralogie et de Cristallographie 98, 11-18, (1975).

[488] E. Galli, A. Alberti, Lithos 6, 83-90, (1973).

[489] A. Alberti, R. Rinaldi, G. Vezzalini, Physics and Chemistry of Minerals 2, 365-375, (1978).

[490] S.A. Miller, J.C. Taylor, Zeolites 5, 7-10, (1985).

[491] F. Pechar, G. Mattern, Crystal Research and Technology 21, 1029-1034, (1986).

[492] E. Galli, A. Alberti, Bulletin de la Societe Francaise de Mineralogie et de Cristallographie 98, 331-340, (1975).

[493] M. Sacerdoti, A. Sani, G. Vezzalini, Microporous and Mesoporous Materials 30, 103-109, (1999).

[494] E. Meneghinello, A. Alberti, G. Cruciani, M. Sacerdoti, G.J. McIntyre, P. Ciambelli, M.T. Rapacciuolo, European Journal of Mineralogy 12, 1123-1129, (2000).

[495] M. Sacerdoti, I. Gomedi, Bulletin de Mineralogie 107, 799-804, (1984).

[496] J.L. Schlenker, J.J. Pluth, J.V. Smith, Acta Crystallographica B 33, 2907-2910, (1977).

[497] A.J. Perrotta, J.V. Smith, Acta Crystallographica 17, 857-862, (1964).

[498] R. Cabellal, G. Lucchetti, A. Palenzona, S. Quartieri, G. Vezzalini, European Journal of Mineralogy 5, 353-360, (1993).

[499] G. Artioli, J.V. Smith, A. Kvick, Acta Crystallographica C 41, 492-497, (1985).

[500] K. Stahl, J.C. Hanson, Microporous and Mesoporous Materials 32, 147-158, (1999).

[501] A. Alberti, M. Sacerdoti, S. Quartieri, G. Vezzalini, Physics and Chemistry of Minerals 26, 181-186, (1999).

[502] H. Ghobarkar, O. Schäf: German Patent, DE 198 24 184 A 1, disclosed 2. 12. 1999.

[503] S. Khodabandeh, G. Lee, M.E. Davis, Microporous Materials 11(1-2), 87-95, (1997).

[504] M.W. Barnes, B.E. Scheetz, Mater. Res. Soc. Symp. Proc. 179(Spec. Cem. Adv. Prop.), 243-272, (1991).

If you have any concerns about our products,
you can contact us on
ProductSafety@springernature.com

In case Publisher is established outside the EU,
the EU authorized representative is:
Springer Nature Customer Service Center GmbH
Europaplatz 3, 69115 Heidelberg, Germany

Printed by Libri Plureos GmbH
in Hamburg, Germany